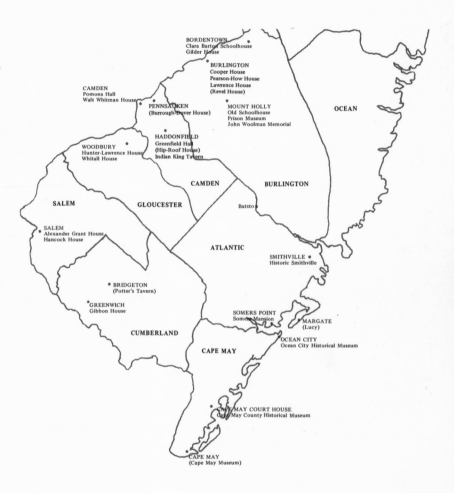

BORDENTOWN
Clara Barton Schoolhouse *
Gilder House

BURLINGTON
Cooper House
Pearson-How House
Lawrence House
(Revel House)

CAMDEN
Pomona Hall
Walt Whitman House

PENNSAUKEN *
(Burrough-Dover House)

MOUNT HOLLY
Old Schoolhouse
Prison Museum
John Woolman Memorial

OCEAN

HADDONFIELD
Greenfield Hall
(Hip-Roof House)
Indian King Tavern

WOODBURY *
Hunter-Lawrence House
Whitall House

CAMDEN

BURLINGTON

SALEM

GLOUCESTER

Batsto

SALEM
* Alexander Grant House
Hancock House

ATLANTIC

SMITHVILLE *
Historic Smithville

* BRIDGETON
(Potter's Tavern)

* GREENWICH
Gibbon House

SOMERS POINT
Somers Mansion

MARGATE
(Lucy)

CUMBERLAND

OCEAN CITY
Ocean City Historical Museum

CAPE MAY

* CAPE MAY COURT HOUSE
Cape May County Historical Museum

* CAPE MAY
(Cape May Museum)

NEW JERSEY'S HISTORIC HOUSES

NEW JERSEY'S HISTORIC HOUSES

A Guide to Homes
Open to the Public

Sibyl McC. Groff

South Brunswick and New York: A. S. Barnes and Co.
London: Thomas Yoseloff Ltd

A. S. Barnes and Co., Inc.
Cranbury, New Jersey 08512

Thomas Yoseloff Ltd
108 New Bond Street
London W1Y OQX, England

ISBN 0-498-07842-6 (cloth)
ISBN 0-498-07878-7 (paper)
Printed in the United States of America

To
SUPER and TILLIE—
for their support of historic houses and me!

CONTENTS

INTRODUCTION

Posters, commercials, travel agents, and newspaper articles implore us to "Discover America." My husband and I have driven around this country to do just that, and sometimes wonder exactly what we are meant to find. Is it crowded highways lined with a plentitude of discount stores, pizza parlors, drive-ins, and gas stations all displaying huge, garish neon signs—a nationwide Route 22? Or is it to discover the turnpikes, the salvation of rapid travel, which bypass the towns and villages whose buildings provide a living testimony to our heritage? We soon realized that by searching, and by chance, one could find architectural vestiges of that heritage. But we also learned that in many instances these historic reminders of the past were threatened by destruction in order to make way for bigger and supposedly better roads, buildings, and parking lots.

Driving along we were reminded of when we had lived in England and were fortunate enough to travel a great deal through the countryside. Our constant companion was an annual magazine, *Historic Houses, Castles and Gardens,* which gave brief descriptions of the houses and their contents, opening hours, admission fees, driving directions, and maps. With the help of this inexpensive publication, we visited many historic houses in England and

Scotland. Seeing how people lived in different centuries made history, architecture, and the decorative arts come alive, thereby providing us with a greater understanding of the country and its people.

Having discovered that there are many historic houses in this country, some of which face an uncertain future, and recalling the value of the English guidebook, I decided that this country needed a similar guide to historic houses that are open to the public. Living in New Jersey, I started here. So now I present to you the story of New Jersey as told through the medium of its historic houses. It should be stressed that I am not an expert on history, architecture, or antiques; for that matter I am not even a professional writer. My intent is to interest readers in seeing and preserving their American heritage. I wish to express my apologies if I have used a misnomer in describing an antique, architectural style, or historical event, or if I have not stressed what a historian, antiquarian, or architect feels is significant in his field. I should also like to point out that my emphasis in these write-ups is the furniture, which is what I most appreciate and enjoy, and the field in which I have received most of my training.

Now that I have explained my motivation in writing the guide, why do I feel that you should visit these houses?

First, as I mentioned above with our experience in England, there is the educational value and sense of history that can be gained from visiting houses as you trace our country's development through the centuries. However, in contrast to the "stately homes" of England, which was already a developed and prosperous nation, remember that the United States in the late seventeenth and eighteenth centuries was a new country, scarcely beyond the frontier stage. Thus the homes were small and contained only the simplest furniture and bare necessities. Beginning with these earlier days, it is interesting to follow the development of house construction, architectural design both inside and outside, domestic equipment and utensils, and

furniture and fittings. You should then better appreciate
how much we take for granted in this modern day of "push-
button" comfort and automation.

Secondly, I hope that you will become interested in old
houses and their cause. Join and support a local historical
society, buy an old house, or join a committee to oppose
the town council when it decides to tear down a late-eigh-
teenth-century house where the ubiquitous George Washington slept. There are not that many historic, Early American houses left in this country, and it is a sad thought
that once they are gone they cannot be replaced. As you
will see in this book, a great many of the historic houses
are owned and maintained by local or county organizations
or groups of private citizens, usually called historical societies. These groups have either raised the money to
purchase a historic house or have been left a historic house
by an estate. Many of these houses when bought or received
were in a dilapidated condition and required extensive
and expensive restoration. Once restored, the maintenance
and furnishing is dear. So if you have a local historical society please join and contribute your time and gifts of
furnishings or financial aid. In return you will probably
learn a great deal about the history of your community
and state. In addition, many historical societies sponsor
interesting events, such as antique shows, lectures, special
exhibits, and tours of private historic homes that are not
usually open to the public.

Thirdly, some of these historic houses are owned by the
federal, state, or county government, and we as taxpayers
have the right to voice our opinion that the respective gov-
ernments should play a bigger role in obtaining and main-
taining historic houses. Albeit there are many demands on
the various government budgets, unless we write our U.S.
Congressman or Senator, or state or local representatives,
urging them to further the cause of historic preservation,
the situation will never improve. In this guide there are
numerous properties described that are owned and main-

tained by the State of New Jersey, and some of these historic houses badly need repair or paint. However, the
amount allocated to the Historic Sites budget isn't anywhere near the amount required to maintain the houses
and contents properly. So if you visit a government-owned
property that you think bady needs a paint job (or other refurbishing), drop a line to some government officials. It
can't do any harm and may just result in paint flowing.

As I traveled through New Jersey, I was most impressed
by the number of active local and county organizations. It
appears that people are awakening to the importance of
historic houses and the need to preserve them. In the year
that I have been compiling this guide there have been
many exciting developments. To mention some, the Monmouth County Historical Association is restoring the pre-
Revolutionary Allen House in Shrewsbury as a tavern. (An
archeological survey of the property located pieces of salt
glaze, pottery, and bottles. The building was the "Sign of
the Blue Ball Tavern" from 1754–1774.) The Hunterdon
County Historical Society in Flemington purchased the
Greek Revival style "Doric" house and is restoring it. The
Pennsauken Historical Society is restoring the eighteenth-
century Burrough-Dover House. In Bridgeton progress is
being made on the restoration of the Potter's Tavern,
which dates back to 1740. In Lambertville the State is restoring the Marshall house, the childhood home of James
Marshall, one of the discoverers of gold in California in
1849. In Morristown an elegant Victorian house Acorn
Hall with its original furnishings has been given by its
benevolent owner to the Morris County Historical Society.
Let's hope that more follow this praiseworthy and generous
example! And finally, as though to demonstrate that there
is some whimsy in restoration work, in Margate "Lucy," a
65-foot elephant that was built in 1882 of wood and tin, has
just been saved from demolition. This architectural folly
with 13 rooms will become a children's library. Incidentally, it is possible that some of these historic properties

are now open, so check with them if you are in their area.

Please accept my apologies if I have neglected to mention a historic house in your vicinity that has been saved, donated, restored, or opened, for there are so many exciting and mostly encouraging announcements in the historic preservation field these days that it is difficult to keep abreast of all.

I close by thanking all who so generously provided me with information and support for this guide and inspired me to "discover New Jersey."

HOW TO USE THE GUIDE

In order to facilitate your use of this guide, I have prepared some comments that you should consider before setting out.

Description of Some Towns: You perhaps will wonder why I have described some localities and not others. The answer is that I have only described communities that are historically, architecturally, or aesthetically noteworthy that I feel are relatively unknown. Furthermore, if the historic house in the locality is an integral part of the atmosphere and complexion of the community, I have attempted to point this out.

Descriptions of the Houses: I have tried to present both a historical and architectural account of the houses, followed by a description of some of their contents, especially the furniture. I have indicated in the text those that I have found to be of the greatest interest.

Times of Admission: Many houses discussed in the text are open by appointment only or on a very limited basis. However, don't be dismayed; if you write or call ahead the caretakers of most houses will gladly try to accommodate you at a mutually convenient time. The opening hours have been provided by the houses. They are subject to change, however, since many of the houses rely on volunteers. Therefore, if you plan a lengthy trip it may be advisable to write or telephone in advance to confirm the opening hours.

Groups: It is imperative that groups (approximately ten or more people) make arrangements in advance. This is necessary since so many houses have limited staffs, and a group will often necessitate additional guides. Also, many of the houses are small. Just think what would happen if five busloads arrived at once? You would all be miserable waiting to get in. Furthermore, some of the houses offer groups a presentation as well as the tour, so you can appreciate the need for advance warning. I cannot emphasize enough how important it is for groups to book ahead. Admission fees for groups (especially those from schools) are often reduced or waived.

(R): Means that rest rooms are available.

Admission: If you should see a donation box, please contribute, as most of these historic houses desperately need money. Also, buy some souvenirs, as the proceeds accrue to the house. Notepaper, plates, or books on the house or county make marvelous and unusual gifts.

Times for Tours: I have tried to give an approximation of how long it will take you to visit each historic house. If anything, I feel that my estimate is on the low side and does not take into account dwelling on one thing (i.e., a collector of Early American craftsmen's tools obviously will want to examine closely an extensive collection).

Recreation: In some cases I have pointed out parks and picnicking and recreational facilities. This is by no means a complete list, as I have relied on information obtained from the caretakers.

NEW JERSEY'S HISTORIC HOUSES

ALLAIRE

The Deserted Village of Allaire (c. 1750 with later buildings until 1840)
Allaire State Park

Records indicate that there was a grist mill on this site in 1750. Between 1790–1820 the first ironworks were established and expanded because of the then-prosperous bog ore (limonite) industry.

In 1822 James P. Allaire, a ship engine builder from New York, bought the ironworks, probably because he required the bog ore for his marine foundry in New York. Under Mr. Allaire's leadership business flourished with the production of such items as hollow-ware pots, pans, stoves, kettles, hinges, and sadirons (the predecessors of our modern hand irons). Numerous buildings were constructed so that Allaire became virtually a self-contained community.

After 1846 Allaire's forge and furnace stopped because of the deterioration of the bog iron industry. With the decline of that industry, Allaire slowly decayed and became "the Deserted Village." However, it wasn't until 1957 that a non-profit corporation was established to restore the village as it was in the heyday of James Allaire and the bog iron industry.

An aerial view of the Deserted Village of Allaire shows the remaining buildings of this once-thriving bog iron community. In the lower left you can see the beehive stack, a symbol of Allaire's bygone iron mill days. This historic site and park are administered by the State of New Jersey. (Courtesy New Jersey Department of Environmental Protection, hereafter abbreviated as N.J.D.O.E.P.)

There are four historic houses in the Village:

Row House (c. 1830). As you enter the Village the information center and snack bar are located in this two-story brick building where the workmen and their families resided. Numerous "Row Houses" existed in the Village, and this is the only one standing.

Foreman's Cottage (c. 1827). This has the distinction of being the first brick building made during Mr. Allaire's time. This tiny one-and-a-half-story house (with one room on each floor) was the home of the works foreman and is now furnished with country pieces of that time. It also serves as an operating post office on weekends.

Farm House (c. 1750). This one-and-a-half-story mustard-colored frame house is the oldest building in the Village. During the time of Allaire it probably served as the works manager's house. It has been completely restored, and the five rooms viewed are furnished with Early American antiques. The large fireplace with its separate baking oven and wood bin is particularly handsome.

The Homestead (c. 1750) *and Boarding House* (c. 1835). This two-and-a-half-story frame house dates from

The Foreman's Cottage (c. 1827) once served as the home of Allaire's works foreman and is now a U.S. Post Office. (Courtesy N.J.D.O.E.P.)

This Farm House (c. 1750) is the oldest remaining building at Allaire and is furnished with Early American antiques. (Courtesy N.J.D.O.E.P.)

about 1750. The three-story brick addition was constructed in 1835 as living quarters for the single working men. Today these are both ramshackled, but it is hoped that sufficient funds will be found for their restoration. As they stand now, the Homestead and Boarding House provide a striking contrast to the restored portion of the Village. Hopefully visitors will realize how run down the Village was before restoration began and, therefore, appreciate the time, effort, and money required in restoration work.

Among other buildings visited are a bakery, blacksmith shop, general store, and enameling furnace. These buildings date from Allaire's time as does the church (even though some older timbers were used in building it). Be sure to see the brick stack, which is all that remains of the furnace. The grounds at Allaire are lovely, graced with sycamore trees and a duck pond. A visit to this old restored village is recommended.

DRIVING DIRECTIONS: Take Exit 96 or 97 on the Garden State Parkway, and then head south on Route 34 a short distance to a traffic circle, where you head west on Route 524. The entrance to Allaire State Park will be on your left. Well-placed signs along the way are indeed helpful.

Open: Week before Memorial Day until Labor Day.

          Daily 10 A.M.–5 P.M.

          Sunday 12 P.M.–5 P.M.

          March 15th until week before Memorial Day, and after Labor Day until December 14th.

          Tuesday–Saturday 10 A.M.–5 P.M.

          Sunday 12 P.M.–5 P.M.

          Closed Monday

          Closed December 15th until March 15th

Admission: Adults—25¢

(R)   Children under 12—free when accompanied by an adult

          Parking—50¢ per car

          School groups—$5.00 per busload

The self-guided tour of the Village is aided by a four-page brochure and map, available at no charge at the information center. The tour lasts about an hour and a half. There are costumed hostesses in the buildings open to the public and demonstrations by a carpenter daily and by a blacksmith daily during the summer. Along the way there are well-placed descriptive maps together with pictures showing buildings that existed during Allaire's time but are no longer standing.

There is a snack bar in the information center. There are many items for sale—postcards, stationery, and an expanded booklet on Allaire; a cookbook and other items are for sale in the bakery; penny candy and many other items are sold in the well-stocked general store; and hand-crafted Early American-design stools, cradles, tables, and chairs are offered for sale at the carpenter's shop.

Pine Creek Railroad

An added treat is a ride on an authentic narrow-gauge railroad. The ride presently goes nearly a mile, and there are plans to expand the railway.

Open: Beginning of May through June, and September and October on weekends:
    12:30 P.M.–5:30 P.M.
    July and August: Daily 12:30 P.M.–5:30 P.M.
    Closed the rest of the year

Admission: 35¢ per person; children under three are free

Note: Allaire State Park provides a wide spectrum of recreational facilities: playgrounds and playfields, picnic tables and fireplaces, fishing on the duck pond (children under 14 only) , a nature center, and four well-marked trails (guided tours daily in the summer and on weekends during the other months) .

ANNANDALE

**Watercress Farm (c. mid-eighteenth century, with addition
c. 1837)
Route 31**

After visiting the grand mansions, castles, and gardens
in England a few years ago, the owner of Watercress Farm
(who had recently retired from the investment field) and
his wife were inspired to open their magnificent gardens
and home to the public.

The Pennsylvania Dutch style farmhouse commands a
lovely setting of glorious gardens, trees, and a pond. As you
face the house, the smaller section to your right is known
to predate 1750, although the exact date has not been
determined. The larger main section, on your left, was
added in 1837. Both sections of the farmhouse are built of
stone covered with stucco, a characteristic of early Pennsylvania Dutch construction. Other additions have been
made since the Goldsmiths, the present owners, bought
the farmhouse in 1936.

Watercress Farm is best known for its gardens, but four
rooms in the house are open to the public. The formal
living room and the dining room are located in the 1837
addition. Fortunately, the original floorboards, fireplaces,

25

Watercress Farm is the residence of Mr. and Mrs. Bertram Goldsmith, who graciously open their house and gardens to the public from April through October. (Courtesy Bertram Goldsmith, photo by The Ann Oakes Studio, Clinton, N.J.)

handsome mantels, and panelling around the recessed windows are intact. The oldest extant portion of the house has been converted into an informal living room with a balcony. Some of the old beams and part of the old fireplace with its beehive oven are seen.

The farmhouse is tastefully furnished with antique furniture, china, and silver as well as with twentieth-century furnishings and items of interest acquired by the Goldsmiths on their world travels. Of interest are four chairs showing the Spanish influence that were brought east in the 1820s when Mrs. Goldsmith's ancestors moved to this area from Texas. Her forebearers also brought the

The living room of Watercress Farm features the blue and white needlepoint rug made by Mrs. Goldsmith. (Courtesy Bertram Goldsmith, photo by The Ann Oakes Studio, Clinton, N.J.)

two family portraits seen in the informal living room. These are examples of portraits that were executed by itinerant artists, who would arrive at one's house with nearly completed canvases with only the face and hands to be filled in.

Needlepoint fanciers will appreciate the beautiful rug that Mrs. Goldsmith stitched, which features 48 different flowers that grow on the farm. There are other needlepoint rugs and seat covers; her beautiful work truly inspired me.

You will enjoy a visit to Watercress Farm, where old and new are harmoniously blended. Incidentally, the thoughtful owners will let you pick free watercress!

DRIVING DIRECTIONS: Driving west on Interstate 78, exit at Annandale. Continue a few blocks through attractive Annandale until you see the large sign, Route 31 South—Flemington, where you go left. Follow Route 31 two miles south until you see the sign for Watercress Farm, on your left. For a landmark, the Farm's driveway is adjacent to a restaurant, "Old Timbers."

Open: Wednesday–Sunday 9:30 A.M.–Sunset April 1–October 31 (Appointments can also be arranged on Mondays and Tuesdays) Advance appointments are advised. All groups by appointment only. *Note:* Since the opening of Watercress Farm depends on the state of the gardens it is advisable to check with Mr. Bertram Goldsmith at 201–735–7010 at the beginning and end of the season to see if they are open.

Admission: Adults: $1.00 (Gardening enthusiasts will be delighted to know that if you purchase $5.00 worth of plants, your admission fee is refunded)
Children under 12: free

The tour of the gardens, greenhouses, and farmhouse is usually conducted by the most interesting and affable owner and lasts about one and one-half hours. It is possible to tour the gardens unescorted but the tour of the house is guided. There are several free pamphlets available. There is a wide range of plants for sale.

BATSTO

**Batsto (c. 1785 with later additions
until 1850s)
R. D. #1, Hammonton**

Driving through this relatively desolate part of New Jersey, referred to as the "Pine Barrens," one is amazed when the tower of Batsto Mansion appears. Batsto Mansion is the manor house of this historic village whose origin dates back to 1766 when Charles Reed began the operation of a bog iron furnace. With plentiful bog iron ore (limonite) and timber for fuel, the village became an important iron maker. The iron works were expanded later and many products were made. Among these were ammunition for the colonies in the Revolutionary War and for the new Republic in the War of 1812, kitchen pots and stoves, pipe for water systems, and firebacks for fireplaces. The bog iron industry flourished here until the 1830s when the discovery of coal (a more economical fuel than wood) in Pennsylvania and advanced iron-making processes doomed the bog iron industry. Batsto's furnaces were closed in 1848.

However, the enterprising owner at that time saw opportunities in glassmaking and began operations in 1846.

This is an aerial view of the historical village of Batsto, showing the mansion and outer buildings, with the pine barrens in the background. Batsto is a New Jersey historic site. (Courtesy N.J.D.O.E.P.)

Among the items made here were bottles, window glass. and glass for the gaslights. The town's second industry was unfortunately not long lived, for the descendants of the owners were uninterested in the glassworks and stopped operations in 1867.

In 1876 Joseph Wharton, a wealthy man who had founded the nickel industry in this country, bought the remains of the then-deserted village and vast acreage in its environs. Wharton, now best known for the graduate business school bearing his name, envisioned damming up the streams and rivers in the area and pumping the water to Philadelphia. His idea was never realized because it was

Batsto Mansion was remodeled in Victorian times, but a portion of it dates from 1785. The impressive tower was built as a fire lookout. (Courtesy N.J.D.O.E.P.)

voted against by the New Jersey legislature. However, he did remodel and expand the mansion and, as he became a gentleman farmer, built numerous outbuildings in the village.

The State acquired the large Wharton tract in 1954 and since then has done a grand job restoring the village in an effort to show the various stages of development in its history.

Numerous buildings are open to the public. Among these are the general store and post office (portions are pre-Revolutionary and later, c. 1847), grist mill (c. 1828), blacksmith shop (c. 1850), and saw mill (c. 1882). However, the mansion, which dominates the village, is our primary concern. The oldest portion of the house dates to 1785, and it has many additions which were built between 1815 and 1878. The present rambling, essentially Victorian style mansion consists of several sections that were added at various times during the village's history. Wharton, for example, completely remodeled the interior and added the tower, porches, and a new facade.

Inside one sees the dining room and two parlors, which are tastefully furnished with antiques ranging from the late-eighteenth to mid-nineteenth century, thereby showing furniture styles during the history of the house. Batsto is most fortunate that several organizations have contributed a wide array of antiques. The kitchen is also on the ground floor and is said to be the oldest section of the house, having been built shortly after the end of the Revolutionary War. It features a Victorian stove and accoutrements and a good collection of ironstone.

Upstairs there are three bedrooms and a child's bedroom that has an interesting display of toys and children's furniture. In the hall the late eighteenth-century blanket chest made of cedar provides a delightful aroma. The decorated brass fixture seen in one of the bedrooms is a bell. Mr. Wharton devised and installed this "bell-buzzer" system where, by ringing the bell, it would buzz in the servant's quarters.

For the more adventuresome, a climb up the many steps to the top of the tower is worthwhile for the magnificent view. (The tower's original purpose was a fire lookout.)

The drawing room of Batsto Mansion has decorative stencilling beneath the molding. (Courtesy N.J.D.O.E.P.)

When leaving the mansion, the visitor should walk down the hill, over the small bridge, and past the sheep grazing in the field to the street flanked by many workmen's cottages. These two-story, weathered, wooden structures consisting of a living room, bedroom, and kitchen downstairs, with more living quarters upstairs, were built in the 1820s. At the far end of the street there are some double attached cottages that date from the 1850s. Interestingly, some of these cottages are occupied by tenants, or their heirs, who lived during Mr. Wharton's time.

The one cottage that is open to the public is painted white, as the others will be when restored. The interior is furnished with typical workmen's country furniture dating from the 1820s and provides a striking and realistic contrast to the elegant furnishings and decor of the mansion. The state is presently restoring the other workmen's cottages where Early American crafts will be demonstrated. The weaver's cottage has been completed, and the resident weaver can be observed busily working on her looms. It is

This rare fireback dates from Colonial times and may be seen in Batsto Mansion. (Courtesy N.J.D.O.E.P.)

hoped that more craftsmen's cottages will soon be operational, as this affords an interesting and educational experience. Other future plans include rebuilding the iron furnace (of which only a small segment of its foundations remains) and establishing a museum at the information center.

At one time this region had many thriving communities due to the bog-iron industry. How fortunate we are that so many buildings in Batsto Village were standing after years of neglect, so that the saga of this village's history can be seen by us today. Don't miss it!

DRIVING DIRECTIONS: Take Exit 50 on the Garden State Parkway, then drive west on Route 542 about 12½ miles. The state has placed many signs pointing the way. You will easily see the Mansion's tower on your right as you approach Batsto.

Open: Memorial Day to Labor Day 10 A.M.–6 P.M.
Rest of the year 11 A.M.–5 P.M., Saturday, Sunday, and Holidays
Summer 11 A.M.–6 P.M., Saturday, Sunday, and Holidays
Closed Christmas and New Year's Day

Admission: Free. One can walk around the village free
(R)   of charge; however, to gain access to the important buildings the guided tour is recommended. Guided tours:
Adults—$1.00
Children 12–18—25¢
Under 12—10¢

The guided tour lasts about one hour and visits the mansion, carriage house, blacksmith shop, and grist mill (in operation on weekends only). Plan on another hour to visit the *workmen's cottage* (open daily at no charge in the summer—weekends otherwise), *weaver's cottage* and perhaps other craft cottages (check at the information center for hours), the *saw mill* (usually operational but closed on Monday), the *ruins of the iron furnace,* the *Post Office* (a working U.S. Post Office) , and *General Store.* The last named serves light refreshments—ice cream and soft drinks —in the summer.

During the summer months the children will love a stage coach ride for 35¢.

At the information center one can obtain free a four-page brochure giving a brief history of Batsto and a map showing the location of the 23 buildings and sites. Postcards, notepaper, expanded books on Batsto and the area, and souvenirs are for sale.

Note: Wharton State Forest offers a wide spectrum of recreational activities—swimming, boating, hiking (with marked trails) , nature trails (guided tours can be arranged) , hunting, fishing, camping (permit required) , cabins (for rent) , canoeing, horseback riding, and picnicking. More information is available at the Batsto Information Center.

BORDENTOWN

Historic Bordentown has had numerous important people in its citizenry. Thomas Paine, the noted political philosopher, was a resident as was Clara Barton, founder of the American Red Cross, who taught school here. Joseph Bonaparte (Napoleon's brother), who was exiled from Spain, lived here in a splendid castle of which only ruins remain today.

Bordentown is fortunate to have many pre-Revolutionary and post-Revolutionary houses still standing. It appears that the townspeople are rapidly becoming more aware of these historic houses as seen by the great number of restorations.

An enjoyable afternoon can be spent here aided by the brochure "Walking Tour of Historic Bordentown," available at no cost at the Bordentown Historic Society (see Gilder House).

DRIVING DIRECTIONS: Take Exit 7 on the New Jersey Turnpike, then follow Route 206 north a couple of miles to the intersection with Route 130. The two routes merge at that point. Follow Routes 206 and 130 north to the first traffic light, taking care to stay on the right since all exits are from the right-hand lane. Exit at that light and, in effect, turn left and proceed across Routes 206 and 130. The first house on your right will be Gilder House. Clara Barton School House is a few blocks south on your left. The Bordentown School is further down the same road on your left.

Clara Barton School House (c. 1839)
142 Crosswicks Street

This one-room brick schoolhouse dating from 1839 was one of the first free public schools in New Jersey. Clara Barton, who later founded the American Red Cross, taught here from 1852 to 1853. This restored schoolhouse is furnished as it would have been in her time.

Clara Barton, founder of the American Red Cross, taught in this one-room schoolhouse in 1852. It is located in Bordentown. (Courtesy David N. Poinsett)

Gilder House is in historic Bordentown. It is maintained by the Bordentown Historical Society. (Courtesy David N. Poinsett)

Open: By appointment only (but since it is a one-room school it is easy to peek in the windows).

During the school year, guides are provided by the Junior School in Bordentown. To arrange a tour write to the Administrative Assistant, Bordentown City Schools, Crosswicks Street, Bordentown, New Jersey. During the summer, contact the Bordentown Historical Society at Gilder House.

Gilder House (c. 1740 with addition c. 1754)
Crosswicks Street

As you face this white clapboard house, the smaller section to your right dates from about 1740. As the family prospered and expanded, the larger section to your left was built.

Inside, one visits only the first floor of the later portion of the house, which contains the entrance hall and two parlors (the latter having their original fireplaces), around which there is handsome panelling. The Bordentown Historical Society maintains this portion of the house.

The front parlor is handsomely furnished with antiques. The grandfather's clock, dating from 1731, was made locally by Samuel Chard. Two tables, dating from the second quarter of the nineteenth century, were made here for Joseph Bonaparte by his French craftsman, John Bouvier. (Yes, that is the great-grandfather of Mrs. Jacqueline Kennedy Onassis!)

Open: By appointment only. (However, there is a resident caretaker, and if you are lucky as I was you'll find him home and he will gladly show you around.)

Admission: Free

The tour is guided and lasts about ten minutes. The "Walking Tour of Bordentown" is available free. There is a park behind the house with facilities for picnicking and grilling as well as playgrounds.

BURLINGTON

In Colonial times Burlington was a thriving port and was also important as the capital of the Province of West Jersey. Some of the town's buildings from that era are still standing. Old St. Mary's Church, dating from 1703, is thought to be the oldest Episcopal Church in the state. Adjacent to it is the new St. Mary's Church (built in 1846 by the well-known architect Richard Upjohn). These churches, together with other noteworthy buildings, are mentioned in a "Walking Tour of Historic Burlington," which is available at no charge at the Burlington County Historical Society Headquarters at Cooper House.

James Fenimore Cooper House (mid-1750s)
457 High Street

James Fenimore Cooper was born here in 1789 and lived his first 14 months in this Colonial dwelling, which his parents rented. The contents of the house are not devoted solely to the novelist but also to exhibits of dolls, china, pottery, and local artifacts of the late eighteenth and early nineteenth centuries. Upstairs is an interesting bedroom with furniture in the French Empire style that belonged to Napoleon's brother, Joseph Bonaparte, who lived in nearby Bordentown for some time.

A noted American was born in each of these two attached houses. James Fenimore Cooper, the novelist, was born in the house on the left in 1789. The house on the right is where James Lawrence, the celebrated Naval Captain of the War of 1812, was born in 1781. (Courtesy N.J.D.-O.E.P.)

Pearson-How House has recently been carefully restored by the Burlington County Historical Society, which also maintains the adjacent James Fenimore Cooper House. (Author's own photograph)

Pearson-How House is also owned by the Burlington County Historical Society and is next door to Cooper House. Tours of the two are combined.

Pearson-How House (c. 1705, addition c. 1725)
455 High Street

This house, situated to the left of Cooper House, has been carefully restored recently and is now open to the public. It is a frame house with a brick facade and has some of its original chair rails, floors, fireplaces, and built-in cupboards.

In order to help us laymen appreciate what is involved in restoring such a building, the Society has done two interesting things. In one of the bedrooms upstairs a section of an interior wall has been left exposed to display the composition of the wall—a mixture of plaster, deer hair, and twigs. Downstairs one can see a small section on a wall where many layers of paint were painstakingly removed in order to determine the original color of the room.

The Society is presently furnishing the dwelling with pieces from the first half of the eighteenth century. Among the items already on display is a Chippendale style chair that belonged to the William Penn family. Although Pearson-How House is sparsely furnished at present, we know that this active Society will acquire the rest of the furnishings in good time.

DRIVING DIRECTIONS: Take Exit 5 on the New Jersey Turnpike. Then drive west approximately four miles on Route 541, which becomes High Street. The houses are on your right.

Open: Sunday 2–5 P.M., or by appointment
Admission: Free
(R)

The tour of the two houses is guided and takes about one hour. Two other buildings are visited on the tour, the library and The Wolcott Museum. The latter has displays of kitchen and farm implements, clothes and accessories from Colonial times, and other rotating exhibits. The four-page walking tour of Burlington and a descriptive write-up of the Society's properties are available free. Booklets on Burlington and souvenir items are for sale.

James Lawrence House (c. 1742, with later additions)
459 High Street

The famous Navy Captain of the War of 1812, James Lawrence, was born here in 1781. His dying cry on the deck of the *Chesapeake,* enshrined in Naval history, was "Don't give up the ship."

The front portion of the house dates from 1742 and the back section was added about 1767. The front of the house was remodeled about 1820 and the side porch was built on. Still later the house was stuccoed.

The four rooms visited contain memorabilia of Captain Lawrence and some period furnishings. The Windsor chairs in the living room belonged to the Lawrence family. An old medical chest dating from the War of 1812 is also of interest. The bottles in the chest contain some old spices and drugs used for medication.

The interior and exterior of the house, which is a New Jersey Historic Site, need painting and restoration.

DRIVING DIRECTIONS: Same as for the Cooper House. The two houses are attached.

Open: Tuesday—Saturday 10 A.M.—12 P.M., 1 P.M. —5 P.M.
            Sunday 2 P.M.—5 P.M.
            Closed Monday, Thanksgiving, Christmas, and New Year's Day
Admission: Free
       (R)
The tour is guided and lasts about ten minutes. A four-page brochure is available free, and there are books on

Thomas Revel House is located in Burlington and is believed to be the oldest extant house in the county. It was
recently saved from demolition and moved to this location.
It is now being restored. (Author's own photograph)

New Jersey, postcards of the house, and other souvenirs
for sale.

Thomas Revel House (c. 1685, with addition c. 1730)
Adjacent to 217 Wood Street

This is considered to be the oldest house in Burlington
County, and it dates from 1685. The second story with its
gambrel roof was added in 1730. This brick house was
recently rescued from the wrecking crew and moved to its
present site. Some of the original panelling and two fire-

places with mantels remain. Plans call for Revel House to be opened in the near future. Let's hope so!

DRIVING DIRECTIONS: From Cooper and Lawrence houses, continue on High Street toward the Delaware River for one block, where you turn left on West Broad Street. You then take your next right onto Wood Street. The house is on your right in the second block.

Open: As Revel House has not been officially re-opened, it is advisable to write ahead for an appointment.

CALDWELL

Grover Cleveland Birthplace (c. 1832)
207 Bloomfield Avenue

The man destined to become both the 22nd and 24th President of the United States was born in 1837 in this early Victorian manse of the First Presbyterian Church where his father, the Reverend Michael Cleveland, was minister. Grover Cleveland is the only President born in the State of New Jersey, and he served two non-consecutive terms, from 1885 to 1889 and 1893 to 1897.

The manse contains memorabilia of the President in the four rooms open to the public. The bedroom where he was born has his Early American cradle. Some of the other pieces of furniture displayed are his desk (used in his Buffalo, New York, law office) and a Victorian-style oak chair that is a combination swivel and rocker that he used in the White House.

DRIVING DIRECTIONS: Take Exit 148 on the Garden State Parkway, then drive west about five and a half miles on Route 506, which is Bloomfield Avenue. The manse will be on your right.

47

In Caldwell one can visit the birthplace of Grover Cleveland, the only President of the United States born in New Jersey. It contains memorabilia of the President and is a New Jersey Historic site. (Courtesy N.J.D.O.E.P.)

Open: Tuesday—Saturday 10 A.M.—12 PM., 1 P.M. —5 P.M.
Sunday 2 P.M.–5 P.M.
Closed Monday, Thanksgiving, Christmas, and New Year's Day
Admission: Adults—25¢
(R)
Children under 12—free

The tour is guided and lasts about 15 minutes. A four-page brochure on the manse and Grover Cleveland is available free. Postcards and notepaper of the manse, books on New Jersey, and souvenirs are for sale. Picnicking is in nearby Grover Cleveland Park.

CAMDEN

Pomona Hall (c. 1726, later addition c. 1788)
Euclid Avenue & Park Boulevard

As you stand facing Pomona Hall, the section to your left was built in 1726 and that to your right in 1788. You can see where the addition was made by the differing texture of the Flemish bond (grey and black brickwork). Also, the older section was initially taller and was cut down when the addition and alterations were made. This can be seen by looking up at the chimney breast, where only part of the initial "C" (for the owner, Cooper) remains. As was customary, the initials of the owners and dates of the two parts of the house are laid in the brickwork in the chimneys on each side of the house.

Inside, the living room (to your left) was originally the kitchen and, being of an earlier period, has a large fireplace. Here you will find some handsome Windsor chairs of pre-revolutionary vintage. The dining room (to your right), with its original, smaller fireplace and built-in cupboard, features a beauty of a sideboard, Hepplewhite in style and made in Philadelphia. Actually, all the antiques in Pomona Hall are quality pieces.

The pine-panelled staircase and partition behind it of

The Camden County Historical Society owns and maintains Pomona Hall in Camden. (Courtesy Camden County Historical Society)

tongue and grooved pine planks are noteworthy. (Don't miss the carved-out heart in the partition.)

The "new" kitchen of 1788 has the appointments of an Early American kitchen. The upstairs of the house is not open now but the Camden County Historical Society plans to furnish these rooms in the Victorian style. Let's hope that this will be accomplished soon.

DRIVING DIRECTIONS: Take Exit 4 on New Jersey Turnpike. Then drive west on Route 73 a very short distance until the intersection of Interstate 295, which you take south about four miles. You exit at Route 30 (White Horse Pike), which you take northwest about two miles around a traffic circle, under a railroad overpass, until it dead ends on Haddon Avenue. You go left on Haddon Avenue past the Harleigh Cemetery (where Walt Whitman is buried) a few blocks to Euclid Avenue. You go right and will shortly see Pomona Hall on your right at the corner of Park Boulevard.

Open: Tuesday–Friday 12:30–4:30 P.M.
                Sunday 12:30–4:30 P.M.

Admission: Free
     (R)

The guided tour lasts about one-half hour. Postcards of the house and historical publications about Camden and the county are for sale.

Note: A new building adjacent to Pomona Hall houses an

This handsome staircase is one of the outstanding features of Pomona Hall and should not be missed. (Courtesy Camden County Historical Society)

excellent library for the history of the state and county and genealogical material. Another two-story building behind the Hall has exhibits of Revolutionary and Civil War memorabilia, Indian relics, lighting, and lamps. This building also has an interesting row of shops with displays of Colonial crafts and tools, e.g., coopersmith, blacksmith, wheelwright and saddler's shops. Pomona Hall and the other Camden County Historical buildings are interesting and educational.

Walt Whitman House (c. 1848)
330 Mickle Street

Whitman lived in Camden from 1873 until his death in 1892 and resided in this Victorian "row house" during his last eight years. The tour starts in the parlor, where one sees the poet's possessions such as his favorite Victorian chair (which he received for Christmas in 1884) and display cases filled with Whitman memorabilia. On your way upstairs take note of his death notice framed on the wall.

Upstairs is his bedroom with the oak bed that was given to the ailing poet by his friends in 1891. On both floors of this small house are copies of Whitman's books, letters, and manuscripts.

The architecture and furnishings are modest but admirers of Whitman should enjoy a visit if only to observe the rather simple surroundings of his last years.

DRIVING DIRECTIONS: Take Exit 4 on the New Jersey Turnpike. Then drive west on Route 73 about one and a half miles to Route 38 south. Approximately six miles toward the center of Camden, Route 38 meets Route 130 (Crescent Boulevard). You go straight on what becomes Kaign Avenue a short distance and then right on Haddon Avenue. Be prepared to take a left on Benson Street shortly after seeing a sign "Broadway Business District."

This "row house" dating from the mid-nineteenth century is located in Camden. Walt Whitman, the poet, spent his later years here. (Courtesy N.J.D.O.E.P.)

Follow this a few blocks to 3rd Street where you go right and then right again on Mickle Street, where the house is on your right, #330. (These directions are rather complicated, but downtown Camden is in the midst of an urban renewal program and many streets

Walt Whitman's bedroom has a tub under the bed that was used for bathing the ailing poet. (Courtesy N.J.D.O.E.P.)

are closed. It is, therefore, advisable to stop at a gas station and recheck to insure that these streets are still open.)

Open: Tuesday—Saturday 10 A.M.—12 P.M., 1 P.M.
                —5P.M.
                Sunday 2 P.M.–5 P.M.
                Closed Monday, Thanksgiving, Christmas,
                and New Year's Day

Admission: Adults—25¢
        (R) Children under 12—free

The guided tour lasts about 20 minutes. There is a booklet on Whitman that may be purchased for 25¢. Postcards are also for sale.

CAPE MAY COURT HOUSE

This small town should not be confused with Cape May, which lies about ten miles south and is discussed below.

Cape May County Historical Museum
Cape May County Court House

This museum is located in the basement of the County Court House and is included because it contains a room dating from about 1790, saved from a post-Revolutionary house that was being demolished. This room, noted for its original stencilling on the walls, is furnished with a variety of antiques such as a 200-year-old dower chest and some Pennsylvania Dutch Windsor chairs.

The Colonial kitchen contains a hand-carved mantle and panelling dating from 1781. The former has been placed over and around a new fireplace built of old bricks to the original dimensions.

In the Country General Store one can also see how people did their shopping in the late nineteenth century. It is equipped with a coffee grinder, pork barrel, and other items.

The other displays help tell the history of Cape May County with Indian artifacts, Early American craftsmen's tools, ship models, military equipment, and Cape May

This room was moved from a demolished house and re-located within the Cape May County Historical Museum. Dating from about 1790, it has original and unusual stencilling on the walls. (Courtesy David N. Poinsett)

Diamonds (a stone found on the beaches in the vicinity).
DRIVING DIRECTIONS: Use Exit 10 on the Garden State Parkway and then follow the signs to Route 9, which is but a short distance away. Turn left on Route 9 and the Court House will be a short distance on your right. There are many signs to direct you to the museum.

Open: Daily 9 A.M.–4 P.M.
           Closed Sunday and Holidays
Admission: Free

The tour is semi-guided and lasts about 20 minutes. A brochure describing the contents of the museum is available for 10¢. Postcards, stationery, and books on the

This is the Cape May County Court House, located in Cape May Court House. The basement houses the Cape May County Historical Museum. (Courtesy David N. Poinsett)

county and state are for sale. About one mile north is Cape May County Park where picnicking and recreational facilities are available.

CAPE MAY

As mentioned earlier, Cape May is situated about ten miles south of Cape May Court House. Well known as a resort town, Cape May is thought to have the greatest concentration of Victorian buildings in the East. The town is currently restoring some of these as part of its urban renewal program. There is a Victorian museum that is scheduled to be relocated elsewhere in the town. The museum has a collection of furniture, clothes, and memorabilia of this seaside resort. Should you visit Cape May inquire at the pavilion on the beach to find out where the museum has been relocated and if it is open. While there take advantage of the lovely beaches and, in the summer months, take a ride on the Cape May trolley.

When Cape May is finished with its face-lifting it should be a real Victorian gem.

CLINTON

Although it is next to a new Interstate highway, Clinton is a tranquil town. The streets are graced with glistening white houses, old and new. The view from the bridge in the center of town overlooking the waterfall at the convergence of the Spruce Run and the Raritan River is most picturesque with the Old Red Mill to your left and the Old Stone Mill (c. 1756) to your right. (The latter is the Hunterdon County Art Center, which has exhibits and a craft shop. It is open on the same days as the Old Red Mill from 2–5.)

Old Red Mill (c. 1763 with addition c. 1820)

The Old Red Mill contains a working water wheel of eighteenth-century design and several facsimile rooms of the Colonial and Victorian periods. The 1760s are represented in the well-appointed Colonial kitchen, which contains many interesting utensils. The other rooms—a child's bedroom, a parlor, a kitchen, and an adult's bedroom—are furnished in the Victorian style.

Over the years the mill has processed diverse products, such as grain, lime, and linseed oil. It even produced electricity at the turn of this century. The rustic interior pro-

In Clinton one can visit the Old Red Mill in this attractive setting. It was built in 1763 and today houses the interesting Clinton Historical Museum. (Courtesy David N. Poinsett)

vides a fitting setting for the extensive craft exhibits, among these are a carpenter shop, a cobbler shop, and a smithy. There are also exhibits of farm and household

equipment. The displays are arranged to show the stages
of development of many household appliances that we now
take for granted, such as the washing machine. The ambitious Clinton Historical Museum plans to use some older
buildings behind the mill for more displays of crafts.

DRIVING DIRECTIONS: Turn off of Interstate 78 at the
Clinton-Pittstown exit. After leaving the ramp, you will
dead end into Route 173 (Main Street). Turn right
and proceed about one-half mile until you see (on your
left) an old coaching inn with large columns called
The Clinton Inn (c. 1743). Take your next left after
the Inn and, as you round the bend, you will see the
Old Red Mill on your left.

The interior of the Old Red Mill contains several facsimile
rooms like this Victorian kitchen. (Courtesy David N.
Poinsett)

Open: Tuesday—Friday 1 P.M.–5 P.M.
                Saturday and Sunday 10 A.M.–5 P.M.
                Closed Monday and November through
                March
Admission: Adults—50¢
    (R)   Children—25¢

The tour is self-guided (aided by posted placards) and lasts about one hour. Guided tours can be arranged with the informative and vivacious Director by writing in advance. An illustrated brochure is available free and books on New Jersey, postcards, and crafts may be purchased.

The Museum also sponsors concerts and travelogues on some Sunday evenings during the summer. Parents will be relieved to know that a baby sitting service is available while they tour the Old Red Mill.

Picnicking is available at the nearby Spruce Run Park and the helpful Museum staff will gladly give you directions. Be sure to take a boat ride on the river for a nominal fee in a boat owned by the museum.

CRANBURY

Although it is surrounded by ever-increasing industrialization and housing developments, the core of this town remains unspoiled. There are some pre-revolutionary houses in that section but most of the brightly painted white houses and buildings are of mid-nineteenth-century vintage.

DRIVING DIRECTIONS: Take Exit 8A on the New Jersey Turnpike. When leaving the toll booths take a right onto Route 32. Get in the lefthand lane as you will very shortly go left at the first traffic light onto Route 535, which you will follow for about $3\frac{1}{2}$ miles across Route 130 and into Cranbury, where you will dead end on Main Street. Go left on Main Street and continue past the pond about one-half mile. The Cranbury Museum will be on your left behind 15 Main Street.

Cranbury Museum (c. 1860)
15 S. Main Street

This four-room house was recently opened to visitors. The present frame structure, dating from the mid-nineteenth century, was most likely built as workmen's living quarters.

The Cranbury Museum was opened recently. (Author's own photograph)

A living room and a dining room are downstairs and a bedroom and museum room for exhibits are upstairs. Furniture from Victorian times to the turn of this century and memorabilia of Cranbury and its families will be featured. More specifics are not available at present as the Museum is in the process of obtaining the contents for the house. It is a delight to know that this charming town now has a historic house open to the public.

Open: Tuesday and Friday, 2 P.M.–5 P.M.
               Or by appointment
Admission: Free
The tour is semi-guided.

Newold House (c. 1750 with later additions)
18 S. Main Street

While I was taking photographs of the Cranbury Museum, a woman asked if she could be of assistance. Our ensuing conversation was about historic houses and this book. Mrs. Morgan informed me that she did live in a historic house that had occasionally been opened for house tours and she thought that the concept of this book was so worthwhile that she and her husband decided to open their house as well by appointment. Let's hope that more people in the future will want to share their historic houses.

Newold House. The owners of this historic house in Cranbury kindly allow visitors to view their home by appointment. (Courtesy **The Home News** in New Brunswick, photo by Will Gainfort)

The kitchen-family room is in the oldest section of New-old House. Gracious living is possible in old houses. (Cour-tesy **The Home News** in New Brunswick, photo by Will Gainfort)

The earliest section of this residence dates from the mid-eighteenth century and originally consisted of one room downstairs and a loft above. It is now incorporated in the kitchen family room and features the original ceiling beams and a large fireplace with its original crane. The house was expanded several times. The exact dates of the additions are not known but the owners plan to do more research on this matter.

At one point the house was a double house and it now contains 15 rooms. One visits four of the downstairs rooms, which are attractively furnished with some Colonial and later antiques. A military satchel, boots, and medicine bottles dating from Revolutionary times were unearthed from the earliest section during repairs, and these items can be seen also.

Open: By appointment only. Write Mrs. James
               Morgan, 18 S. Main Street

Admission: Free

The tour is guided by the enthusiastic couple and takes about 20 minutes. There are no items pertaining to the house for sale, but you may want to purchase the dried flower arrangements or pressed flower pictures that are made and sold by Mrs. Morgan.

CRANFORD

Cranford Historical Society (c. 1840, with later additions)
124 North Union Avenue

The headquarters of the Cranford Historical Society is in this modest house, the oldest portion of which dates to 1840. The front porch was added in 1868. Three rooms (one built on in 1963) serve as the Society's museum, featuring a pictorial history and memorabilia of the town and its citizenry. There is some furniture. One piece of interest is the mahogany bookkeeper's desk (c. 1840). The Society also has changing exhibits.

DRIVING DIRECTIONS: Take Exit 137 on the Garden State Parkway. Then drive west on Route 28 (North Avenue) about three-quarters of a mile to the second traffic light. At that light turn right onto Springfield Avenue. At the first traffic light, you turn right again onto Union Avenue. The Society is the first building on your right.

Open: Saturday 10 A.M.–2 P.M.

              Sunday 3 P.M.–5 P.M.

Admission: Free

     (R)

This vine-covered house is now the headquarters of the Cranford Historical Society. The house dates from about 1840 with the porch being added in 1868. (Author's own photograph)

The semi-guided tour lasts about ten minutes. Note-paper and a booklet on the history of Cranford ("300 Years at Crane's Ford"—$1.00) may be purchased.

DUMONT

Joost Zabriskie Homestead (c. 1712, with addition c. 1740)
2 Colonial Court

It is exciting that the owners of this Dutch Colonial farmhouse will allow people to tour their residence by appointment. The smaller section (dating from about 1712) consists of a keeping room with a loft. The much larger section was added about 1740 and has a gambrel roof. Both are made of native New Jersey red sandstone.

You enter the house through the older section and find yourself in the keeping room, which has its original ceiling beams as well as the original built-in corner cupboard. You ascend a few stairs to the dining room. The ceiling was plastered when the present owners bought the house. They removed the plaster and the handsome beams with bead molding were uncovered. This room and the living room have Federal period fireplaces, which were most likely added when the house was remodeled in the early nineteenth century.

Only the downstairs is open to visitors. The owners are avid collectors and have on display Early American kitchen equipment and gadgets, bells, bottles, Delft china, and a foot warmer dating to 1750. Our thanks to this couple for sharing their historic residence with others.

This Dutch Colonial farmhouse in Dumont is privately owned but may be visited by appointment. It is known as the Joost Zabriskie Homestead. (Courtesy Joel Altshuler)

The dining room of Joost Zabriskie Homestead has a Federal Mantle, which was installed in the early nineteenth century. (Courtesy Joel Altshuler)

DRIVING DIRECTIONS: Since this is a private residence, driving directions are provided only by the owners upon request.

Open: By appointment only. Please write to the owners, Mr. and Mrs. Joel Altshuler, at the address shown above.

Admission: Free. The tour of the five rooms is conducted by the owners, and lasts about one hour. There is an interesting book for sale on Dumont's heritage written by Mrs. Altshuler and illustrated by her husband.

ELIZABETH

Boxwood Hall (c. 1750)
1073 East Jersey Street

Boxwood Hall was built in 1750 by Samuel Woodruff, then Mayor of Elizabeth. In 1772 it was purchased by Elias Boudinot, a prominent political figure. A lawyer, Boudinot was President (1783) of the Continental Congress. He entertained Washington here (April 23, 1789) when the latter was on his way to New York for his first inauguration. As a youth, Alexander Hamilton was also a frequent house guest. Incidentally, Boudinot conceived the idea of Thanksgiving Day (though it did not become an official national holiday until the following century).

The house is typically Georgian and is tastefully furnished with Colonial and Federal pieces. Although it was used as a home for the aged from 1871 until 1940, the original floors, panelling, and dental molding around the ceiling are intact.

Note: There are several pre-revolutionary houses (not open to the public) in Elizabeth and those interested in a walking tour should consult a map on the wall in Boxwood Hall. The less ambitious should at least walk up to the corner to 1045 East

This Georgian house in Elizabeth, Boxwood Hall, was at one time the home of Elias Boudinot, a well-known American patriot. It is owned and maintained by the state of New Jersey. (Courtesy N.J.D.O.E.P.)

Jersey Street to see the oldest house in the town, Bonnell House, now the headquarters of the New Jersey Sons of the American Revolution (c. 1665). (You may enter the building if you write and make arrangements. There is little of interest inside now but there are plans to furnish the house.) Also, walk across the street and visit Belcher Mansion.

DRIVING DIRECTIONS: Take Exit 13 on the New Jersey Turnpike and follow signs to Routes 1, 9, and 439. At the first traffic light you proceed straight on Route 439 about one mile until it meets Routes 1 and 9 at a traffic circle. You then take Routes 1 and 9 north about one mile. Be sure to be in the right-hand lane when going over the Elizabeth River as you will want to turn right at the next traffic light (all turns are from the right-hand lane). After exiting Routes 1 and 9, turn left (crossing Routes 1 and 9) and proceed straight ahead. This is E. Jersey Street. Belcher House is one block ahead on your left and Boxwood Hall is one-half block further down the road and will be on your right.

George Washington visited Elias Boudinot when on his way to New York for his inauguration. It is very likely that he was entertained in this attractive dining room at Boxwood Hall. (Courtesy N.J.D.O.E.P.)

This is a corner of the sitting room in Boxwood Hall. (Courtesy N.J.D.O.E.P.)

Open: Tuesday—Saturday, 10 A.M.–12 P.M., 1 P.M.–5 P.M.
             Sunday 2 P.M.–5 P.M.
             Closed Monday, Thanksgiving, Christmas and New Year's Day
Admission: Adults—25¢
     (R)   Children under 12—free

This is a guided tour lasting about 30 minutes. Posted outside each of the eight rooms is a description of the room's contents. A four-page brochure on the house is available free and books on New Jersey, notepaper, postcards, and souvenirs may be purchased.

Picnicking facilities are available at Warinanco Park, about five minutes away. Ask the attendant for directions.

Belcher House (c. late seventeenth century, eighteenth century additions) 1046 East Jersey Street

It is thought that this two-story brick house has been the residence of three prominent citizens of the State. John Ogden, one of the founders of Elizabeth in 1662, is believed to have lived here. The house is named for Jonathan Belcher who, while Royal Governor, lived here from 1751 until his death in 1757. Aaron Ogden, a descendent of John, also resided here. A colonel during the Revolutionary War, Ogden became Governor of New Jersey in 1812 and later a renowned lawyer.

The non-profit Elizabethtown Historical Foundation has carefully restored the outside and inside of this noteworthy house. If you look closely at the facade of the building you can see where the addition was made by observing the different texture and color of the brickwork. Fortunately, the interior has many of its original appointments including panelling, fireplaces, floor boards, and a built-in cupboard.

With the exception of one room, the house is furnished in accordance with an inventory taken by Jonathan Belcher in 1757. The parlor is devoted to Aaron Ogden and has furnishings of his period including a signed Duncan Phyfe sofa and chairs.

In the reconstructed kitchen one can see an original fireback dating to 1742. Another item of interest is a Cromwellian leather chair with brass studs that is believed to date to the late seventeenth century. This chair was found when the house was undergoing restoration.

(DRIVING DIRECTIONS: See page 75.)

Belcher House is down the street from Boxwood Hall in Elizabeth. Several occupants of this well-appointed brick residence are well known in New Jersey's history. (Author's own photograph)

Open: Wednesday 9:30 A.M.–12 noon, September
               through May, or by appointment
Admission: Free
       (R)
The tour is guided and takes about 30 minutes. There is
a four-page brochure available free.

FREEHOLD

This is an attractive town. Besides the places mentioned below, see the old courthouse in the center of town (Main and Court Streets), which dates from 1874. It is in the Georgian design, but has been greatly altered. Near West Main Street is St. Peter's Episcopal Church (33 Throckmorton Street), the predecessor of this church dated to 1702. The present white-shingled church dates from the 1750s.

Monmouth County Historical Association Headquarters (1931)
70 Court Street

This attractive, three-story brick dwelling was built in 1931 in the Georgian style. It certainly cannot qualify as a historic house but must be judged as an excellent reproduction that contains an impressive collection of antiques—furniture, porcelain, silver, and paintings of the seventeenth and eighteenth centuries. Particularly, there is a fine representation of antique furniture made in Monmouth County. These items are artfully displayed in period rooms on the first two floors. Be sure to note the painting *Washington at Monmouth* by Emanuel Luetze

The Headquarters of the Monmouth County Historical Association is housed in this brick building, which dates only from 1931. It has a fine collection of antiques, books, and papers relating to the history of the county. (Courtesy Monmouth County Historical Association)

(the artist who also painted *Washington Crossing the Delaware*) and the rare secretary-bookcase (c. 1760) with a carved bust of William Penn in the pediment.

In an upstairs bedroom there is what is believed to be one of the earliest crewel bedhangings in this country, dating from about 1700, which was made for an early Monmouth County family. On the third floor, the junior museum has interesting exhibits of old dolls, toys, and signs, and Indian, Colonial, and Revolutionary artifacts. Adults as well as children will enjoy this.

This is a corner of one of the rooms at the Headquarters of the Monmouth County Historical Association. (Courtesy Monmouth County Historical Association)

This bedroom features beautiful crewel bedhangings that were stitched about 1700. (Courtesy Monmouth County Historical Association)

DRIVING DIRECTIONS: From the New Jersey Turnpike take Exit 8 and drive east about 12 miles on Route 33, which intersects with Main Street. Take a left on Main Street and continue a short way until you turn left on Court Street. The Headquarters is a few blocks on your left, across from the new County Court House and the monument commemorating the Battle of Monmouth, June 28, 1778.

Open: Tuesday–Saturday 11 A.M.–5 P.M.
                Sunday 2 P.M.–5 P.M.
                Closed December 15th–31st, July 15th–31st

Admission: Free

    (R)

A self-guided tour lasts about one hour and is aided by good posted descriptions. One may arrange in advance for a guided tour. Books and pamphlets on Monmouth County and the state are for sale.

Note: A visit here and to the Association's other properties in Holmdel and Middletown (see respective listings) is recommended. This Association has done a splendid job with their properties in all aspects, and a trip to their three houses will prove interesting and educational for all. The Association also has an extensive library dealing with Monmouth County's history and genealogy.

Clinton's Headquarters (c. 1706)
150 West Main Street

The Monmouth County Historical Association is presently restoring this headquarters of the British General Henry Clinton before the Battle of Monmouth, June 28, 1778. Although now not open to the public, it is worthwhile to take a look at this farmhouse dating from 1706 and added on to in the 1750s. For a simple farmhouse it contains some outstanding panelling and built-in cupboards.

The Monmouth County Historical Association is currently restoring this farmhouse, which served as the headquarters of the British General Clinton before the Battle of Monmouth. (Author's own photograph)

DRIVING DIRECTIONS: From the center of town, drive a few
blocks southwest on Main Street. It is the house on
your left at the junction of Route 9.

Open: Check at the Association's Headquarters on
                 Court Street for a progress report. It just
                 may be open.

(opposite)
Interior view of General Clinton's Headquarters in the
process of restoration. The seascape dates to the construction of the house. (Courtesy David N. Poinsett)

GREENWICH

This small town in the southern section of the state is without doubt one of the most charming and tranquil places left in New Jersey. "Ye Greate Street," which was laid out in 1684, still is the main wide thoroughfare and is graced by lovely old trees and by houses representing many American architectural styles from Colonial to Victorian. Noteworthy houses have signs on them indicating when they were built.

Be sure to stop at the Cumberland County Historical Society (see below) and buy their publication, *Fiftieth Anniversary Celebration of the Unveiling of the Tea Monument in Greenwich.* This booklet, prepared in 1958, describes not only the tea-burning that took place on December 22, 1774, by colonists who refused to pay a tax levied on new shipments of tea, but also the houses and churches in the town and the surrounding area. With it you can spend an enjoyable and educational afternoon on your own walking tour of this unspoiled and refreshing town.

Gibbon House (c. 1730)
Main Street

The Cumberland County Historical Society recently

Gibbon House is located in the attractive southern New Jersey town of Greenwich. In less than a year the Cumberland County Historical Society has done a fine job in renovating this house dating from 1730. (Courtesy David N. Poinsett)

acquired this house, dating from about 1730, which is built in the Flemish bond style with a pent roof (overhang between the first and second stories). In less than a year the Society has done an outstanding job in refurbish-

A corner of the living room in Gibbon House. (Courtesy David N. Poinsett)

ing the house. Fortunately, the fireplaces and two handsome built-in cupboards in the living room were intact. Be sure to peek into the first cupboard so that you can see where there was once a window, which was covered up to avoid the glass tax levied on the Colonies before the Revolutionary War. These cupboards contain a fine representation of china—Chelsea, Beleek, Blue Canton, and Brown Staffordshire, to mention a few. Downstairs you will also see the library and a well-fitted Colonial kitchen, which features a large fireplace, part of which has been rebuilt. Upstairs you will visit a child's room filled with old toys and furniture, the master bedroom, a lady's dressing room, and a room devoted to rotating exhibits.

The house is well furnished with antiques from Colonial

through Victorian times; noteworthy among these are a collection of Ware chairs. Maskell Ware started making these ladderback chairs nearby about 1778, and the family business continued until 1935.

DRIVING DIRECTIONS: From the center of Bridgeton take Route 49 west about a mile up the hill past the Potter's Tavern (c. 1740, which is currently being restored and will be open when completed) on your right and the Old Broad Street Church (c. 1792, lovely Georgian architecture) and cemetery on your left. At the end of the cemetery you go left (or south) on West Avenue, taking your next right on to Greenwich Avenue. Follow this road about six miles. Unfortunately, it is poorly marked, but if you should get lost do not hesitate to ask directions. The friendly local folks were indeed most helpful to me.

Open: Saturday and Sunday 2 P.M.–5 P.M. from the first Sunday in April through October or by appointment.

Admission: Adults—25¢
  (R)   12–18—10¢
            Children under 12—free

Publications on the county and postcards and notepaper of the house and Greenwich are for sale. The tour is self-guided and takes about 30 minutes. There are posted descriptions of the contents of each room.

HADDONFIELD

This is a charming town with houses representing many architectural styles from the mid-1700s onward. Should you walk along the main thoroughfare, the King's Highway, you will feel that you are in a town at the end of the eighteenth century (except for the noises of that later invention, the automobile). Besides Greenfield Hall, the Hip-Roof House, and the Indian King Tavern, which are discussed below, many other pre-Revolutionary and post-Revolutionary houses are well preserved and are presently used as private residences or professional offices. How marvelous it is that these old dwellings are still standing and being used.

DRIVING DIRECTIONS: Take Exit 4 on the New Jersey Turnpike. Drive west on Route 73 about one and a half miles and then south on Route 41 (which becomes Route 551) approximately three and a half miles. This, in turn, is the King's Highway. First on your right will be Greenfield Hall and the Hip-Roof House (situated together) and a few blocks further on is the Indian King Tavern.

In pleasing Haddonfield one can visit Greenfield Hall, headquarters of the Haddonfield Historical Society. The small section on the right of the picture is the oldest and was built about 1747. The rest of the house was added about 1841 and today contains many exhibits and antiques. (Author's own photograph)

Greenfield Hall (c. 1747, later additions c. 1841)
Hip-Roof House (c. 1736)
343 King's Highway East

These two totally different houses provide a most striking contrast. The older, modest Hip-Roof House is practically dwarfed by the larger, more stately Greenfield Hall.

Considering the latter first, the miniscule, set-back section to your right is the oldest part of the house. The rest of this three-story brick dwelling was built in 1841 on the foundation of the older house. The more recent portion shows the influence of the Greek Revival style, which can be seen in the columns at the entrance and by the balustrade around the top of the roof.

The house is well preserved and has many of its original appointments, noteworthy among these are the plastered motifs on the ceilings, the carved wooden venetian blind cornices, and the floors. The Haddonfield Historical Society has an interesting and varied collection that is shown in this graceful house. The ground floor has a tastefully furnished keeping room, a Victorian parlor, and two other parlors, one in the front, the other in the rear. The furniture is of the Victorian period or earlier. Two early pieces date from the first quarter of the eighteenth century: a walnut pier table with a marble top and cabriole legs and a looking glass mirror. Both are quality pieces that belonged to Elizabeth Haddon, the namesake of Haddonfield. She came here from England at age 20 to manage her family's properties and resided here until her death in 1762.

There are two furnished bedrooms on the second floor and a room containing displays of dolls, jewelry, and other items, such as memorabilia of Elizabeth Haddon. On view are the certificate of her Quaker marriage to John Estaugh in 1702 and her medicine chest.

The third floor houses a collection of dolls, costumes, and spinning and weaving looms. The basement contains many types of craftsmen's tools, such as those of coppersmiths and cobblers, and domestic equipment, such as a candle dip and mold and a cider press.

Hip-Roof House (c. 1736)

This dear little white clapboard house with its gambrel

roof was moved here several years ago from elsewhere in
Haddonfield. It is thought to be the oldest existing house
in Haddonfield. Elizabeth Haddon Estaugh owned the
house during the last ten years of her life and leased it to
tenants. Detailed studies of the house have been under-
taken, and it is hoped that it will be restored and furnished
in due course. It is not now open to the public.

DRIVING DIRECTIONS: See above under Haddonfield

Open: Tuesday and Thursday 2 P.M.–4:30 P.M. or
             by appointment
             Closed July and August
Admission: Free
      (R)

Adjacent to Greenfield Hall is this tiny house, Hip Roof
House, supposedly the oldest dwelling in Haddonfield. The
Haddonfield Historical Society owns it and is hoping to
restore it soon. (Author's own photograph)

The tour of Greenfield Hall is guided and lasts about 45 minutes. Two five-page pamphlets, "A Sketch of the History of Haddonfield" and "A Word About the Historical Society of Haddonfield (Greenfield Hall and Hip-Roof House)," are available at no charge. Notepaper, tiles, ashtrays, plates of the houses, and other souvenir items are for sale. In the oldest section of Greenfield Hall there is an excellent library on the history of the town and county. It also has genealogical material.

Indian King Tavern (c. 1750)
233 King's Highway East

This old and pleasant-looking inn has been the site of many interesting New Jersey historic events—the most noteworthy being the New Jersey Assembly meeting of May 10, 1777, which accepted the design for the Great Seal of New Jersey, and that of September 20th of the same year, which proclaimed New Jersey a state on all official documents. These meetings took place on the second floor in what was once a ballroom.

This three-story brick structure with painted stucco exterior has a small pent roof extending from above the first floor. Some of the floors and mantels are restored, but the original locks are on the doors.

In the basement the dark wine cellar with its low ceiling is said to have been used during the Revolutionary War to hide either captured enemy soldiers or sought-after Americans. No doubt it was a refreshing place to hide, whatever one's colors.

The three rooms on the ground floor are used as headquarters for three fraternal organizations as is the former ballroom on the second floor. As a result, although they are furnished with some period pieces, the atmosphere of a wayside tavern is missing. However, plans have been drawn up to make one of the rooms into a Colonial taproom, and let's hope that these materialize.

The Indian King Tavern is located in Haddonfield. It is a New Jersey Historic site. (Courtesy N.J.D.O.E.P.)

An interesting piece of furniture in the tavern is a ladderback chair with a rush seat, which dates to the 1750s and belonged to the Matlock family who built this inn.

There are four more rooms on the second floor. One houses an exhibit of Haddonfield pottery, made in this town from about 1805 to 1868. Another room has a collection of late seventeenth and eighteenth-century toys, dolls, and children's furniture, which the younger visitors will enjoy. The third floor is not open to the public.

Open: Tuesday–Saturday 10 A.M.–12 P.M., 1 P.M. –5 P.M.

Sunday 2 P.M.–5 P.M.

Closed Monday, Thanksgiving, Christmas and New Year's Day

Admission: Adults—25¢
    (R)   Children under 12—free

The tour is self-guided with descriptions of each room posted. It lasts about 20 minutes. A four-page brochure on the Tavern is available at no charge. Notepaper, postcards of the Tavern, books on New Jersey, and souvenirs are for sale.

HOLMDEL

Holmes-Hendrickson House (c. 1717)
Longstreet Road

Built about 1717, this Dutch Colonial farmhouse was originally on the site of the nearby Bell laboratories. When the labs were built in 1959, the company moved the house to its existing location and presented it to the Monmouth County Historical Association. Since receiving the house, the Association has painstakingly restored its interior and exterior. The fireplaces and most of the floors and hardware are original, however.

As you enter the house through the kitchen, note the Dutch Beehive oven and rare, curved-back kitchen "settle" or bench. (The latter has never been painted.) The other rooms downstairs are tastefully furnished, mostly with Queen Anne and William and Mary pieces. Of particular interest are the William and Mary day bed with reclining back and the "fold-a-way" bed in the downstairs bedroom.

Note: The Association has not dedicated this house officially since it is still working on restoring and furnishing the upstairs rooms. In fact, when I visited the house, one could see the work in progress upstairs, and portions of the interior walls were ex-

Holmes-Hendrickson House is located in Holmdel and is a Monmouth County Historical Association property. (Courtesy Monmouth County Historical Association)

This is one of the attractively furnished rooms at the Holmes-Hendrickson House. (Courtesy the Monmouth County Historical Association)

The attractive kitchen in the Holmes-Hendrickson House recreates the atmosphere of the mid-eighteenth century. (Courtesy Monmouth County Historical Association)

posed so that their original colors could be determined.

DRIVING DIRECTIONS: Take Exit 114 on the Garden State Parkway. Go southeast a short distance to the first intersection (Crawford's Corner Road), and turn right. Shortly after that, turn left on Longstreet Road, and the house is on your right.

Open: Tuesday and Thursday 1 P.M.–5 P.M.
             May through October, or by appointment.

If you wish to make such an appointment, contact the Monmouth County Historical Society in Freehold.

Admission: Free

Tour is guided and lasts about 30 minutes. Picnicking facilities are available in adjacent Holmdel Park (see also Freehold and Middletown).

HOPEWELL

This is a peaceful town with an attractive main street called Broad. In the town and surrounding countryside you will come upon many attractive dwellings dating from Colonial through Victorian times.

Hopewell Museum (c. 1877)
28 East Broad Street

This Victorian grey house was erected in 1877. Downstairs it has two parlors, one in the Federal period and the other in the Victorian period. Both are appropriately furnished. More antique furniture of the Colonial, Federal, and Victorian periods is on display in the upstairs bedrooms as well as in the upper portion of a new addition. The rest of the house has exhibits of Revolutionary and Civil War military equipment, Colonial kitchen equipment, costumes, musical instruments, and items pertaining to local history. Rotating exhibits are also shown here. The small community of Hopewell must be proud of this museum.

DRIVING DIRECTIONS: Hopewell is located about six miles east of Lambertville on Route 518. As you drive through the town, the House is one and one-half blocks on your left after the traffic light.

The Hopewell Museum is housed in this Victorian edifice.
(Courtesy David N. Poinsett)

Open: Monday, Wednesday and Saturday 2 P.M.–
           5 P.M. or by appointment
Admission: Free
           (R)

The tour is guided and takes about 45 minutes. A brochure on the museum is being prepared, and it is hoped

Victorian times are recreated in this parlor in the Hopewell Museum. (Courtesy David N. Poinsett)

that it will be available for visitors soon. Postcards of the
house and pens are for sale. Picnicking facilities are available in Stoney Brook Park, which is located on Spur Route
518 between the town and Route 31.

LIVINGSTON

Force House (c. 1745, later addition c. 1825)
366 South Livingston Avenue (P. O. Box 200)

When it was built about 1745 this sprawling farmhouse consisted of one room with a loft above. Later, as the Force family prospered and expanded, the three-story addition was erected. The township recently purchased this historic dwelling and the Livingston Historical Society is responsible for its maintenance and furnishings. The interior was in disrepair and the Society has concentrated on refurbishing it. The floors in the newer section have been replaced with old floor boards taken from another house. The original fireplaces were intact and it is interesting to contrast the larger crude one in the older beamed "keeping" room with the smaller one in the newer South Parlor, which possesses a delicately executed Adam-style mantle.

Now that the interior is in better condition, the Society is collecting furnishings. On the first two floors these will date from 1830 or earlier. The main hall has a most unusual Hitchcock bench that is also a rocker and a handsome tall-case clock. The third floor will have a Victorian bedroom and exhibits of old maps and memorabilia of Livingston. Other plans call for the conversion of the non-

The Force House is maintained by the Livingston Historical Society. To the front of this old farmhouse you can see the herb garden in bloom. (Author's own photograph)

vintage garage into a barn for the display of old farm equipment. The Society has already accomplished a lot in a short time.

Condit Williams Cook House (c. 1700)

This little one-room building with a loft was moved to the grounds of the Force House. Dating from about 1700, this wooden structure has been used as a summer cookhouse, a residence, and more recently as a playhouse for local children. Since being saved from the wrecker's ball and moved here, it has been painted a cheery white and

On the grounds of the Force Homestead is this dear little house, Condit Williams Cook House, which dates from 1700. It was saved from demolition and moved here. (Author's own photograph)

the fireplace has been rebuilt with old bricks. It now contains a collection of Early American kitchen utensils and tools. Climb up the steep narrow stairs and look at the loft!

DRIVING DIRECTIONS: Take Exit 144 on the Garden State Parkway and follow the numerous signs to South Orange Avenue (Route 510), which you take west about 7 miles. (On the way you will pass the South Mountain Reservation so plan ahead for a picnic, barbecue, hike, or a visit to the zoo.) You then head north on the John F. Kennedy Memorial Parkway (which becomes Living-

ston Avenue) for about one and three-quarters miles.
The Force House and Condit Williams Cook House will
be on your left, within the Livingston Civic Center.

Open: The second and fourth Sundays during the
               months of September, October, November,
               April, May, and June, 2–4 P.M. Also open
               by appointment.

Admission: Free

   (R)

The tour is guided and lasts about 30 minutes. There is
a seven-page pamphlet on Force House and its history
available free. Notepaper of the house is for sale.

MIDDLETOWN

Settled about 1665, Middletown is one of the oldest towns in Monmouth County. Today, its tree-lined main street, the King's Highway, has houses and public buildings along it that present a scene of the town's earlier days. A walk or drive down the King's Highway is recommended. While doing so, you will wish to take note of the First Baptist Church (chartered in 1688) completed in 1832 and the architecturally simple but attractive Christ Church (chartered in 1702) whose present edifice dates to 1836. Local legend holds that Captain Kidd worshipped in the latter's earlier building.

Marlpit Hall (c. 1684, additions c. 1712)
137 King's Highway

You enter this Dutch Colonial farmhouse in its original core, which dates to 1684. This room, with its original beams and fireplace, served as living room, kitchen, and dining room and is furnished with antiques of the period including a table from 1690, Delft ware, and kitchen equipment. (The original bedroom was behind and is not open to the public.)

In the early part of the eighteenth century (c. 1712) a

wing was added to the house, and this is visited on the second half of the tour. This wing contains three rooms and a hall downstairs and additional bedrooms and a store-room on the second floor. The panelling and woodwork are original, and noteworthy for their craftmanship. The parlor has a magnificent hand-carved, built-in corner cup-board with a sunburst motif inside. Also in the parlor is a fireplace with handsome pilastered panelling. In the hall there is a well-executed Dutch kas, or cupboard, decorated *en grisaille,* painted in shades of grey. The other furni-ture is of the Queen Anne and William and Mary periods.

In the very old and pleasing town of Middletown, one can visit Marlpit Hall, a Dutch Colonial farmhouse. The small section to your right dates to 1684 and the larger portion was added about 1712. (Courtesy Monmouth County Historical Association)

The keeping room is in the oldest section of Marlpit Hall.
(Courtesy Monmouth County Historical Association)

As you leave the house through the newer, massive front
door, notice the original latch and lock, and the rare
bull's eye at the top of the door. Some of the outside of
the house, notably some of the shingles and roof, have been
restored.

Note: Marlpit Hall is owned by the active Monmouth
County Historical Association, which also owns the
Holmes-Hendrickson House in nearby Holmdel.
A visit to that house and to the Association's headquarters in Freehold is recommended. The attendant at Marlpit Hall will gladly give directions to
both. (See Holmdel and Freehold.)

DRIVING DIRECTIONS: Take Exit 114 on the Garden State

Parkway. Head east on Red Hill Road for about two miles. At the end of that road, bear left onto the King's Highway. Marlpit Hall is #137 on your right.

Open: Tuesday, Thursday, Saturday 11 A.M.–5
               P.M.

               Sunday 2 P.M.–5 P.M. or by appointment

Admission: Free

     (R)

The tour of the house is guided and takes 30 minutes. Books and pamphlets about Monmouth County, post-cards and notepaper showing the house, and other items are for sale.

MONTCLAIR

Israel Crane House (c. 1796)
110 Orange Road

This wooden house was built in 1796 by a most enterprising young gentleman, Israel Crane, later in life to be
called "King." He started many successful companies and
ventures during his entrepreneurial career, such as a toll
road between Newark and Caldwell.

Structural studies of this house indicate that initially
it consisted of two stories with a pitch roof. Around 1840
the house was remodelled; the roof was raised and a
straight roof built, around which the classic-style cornice
was added. The cornice contains some most interesting cast
iron grills. Further evidence of the Greek Revival influence can be seen in the treatment of the columns at the
front entrance. Be sure to notice the curious window
frames appended to each side of the house on the second
story. Why these windowless frames were put there is uncertain.

This house was saved from demolition and moved here
by the Montclair Historical Society in August 1965. This
essentially Federal style house possesses some of its original
features, including floors, some window panes, the upstairs

This appealing, essentially Federal style dwelling is known as Crane House and is owned and maintained by the Montclair Historical Society. (Courtesy David N. Poinsett)

fireplaces, and attractive recessed molding around the ceilings. An interesting interior feature is the existence of two staircases, one grander one leading up from the spa-

cious center hall and the other going up from the warm-
ing kitchen.

Various periods are well illustrated with such rooms as a

Crane House has a handsome Federal dining room. (Courtesy David N. Poinsett)

Colonial parlor, Federal dining room, Empire parlor, and an Early American kitchen (the last a reconstruction attached to the house by a passageway). The three bedrooms upstairs are furnished with late eighteenth- and early nineteenth-century pieces. In one of these rooms one can see the exposed wall showing the basic structure of the outer wall called nubbing. Another bedroom is used for exhibits that are changed every month or so.

In a short time the Montclair Historical Society has done a remarkable job refurbishing and furnishing the Crane House, and further improvements and expansion can be expected of this active society. A visit to this well-appointed Federal house is recommended.

DRIVING DIRECTIONS: Take Exit 148 on the Garden State Parkway, then follow directions to Bloomfield Avenue. After turning right onto Bloomfield Avenue ("King" Crane's old toll road), proceed two and a quarter miles, then turn left onto Orange Road. The house is about half a mile down the road, and will be on your right.

Open: Sunday 2 P.M.–5 P.M. or by appointment.
            Closed during July and August

Admission: Free
            (R)

The semi-guided tour lasts about 30 minutes. There are guides as well as posted descriptive placards.

MORRISTOWN

Macculloch Hall (c. 1810)
45 Macculloch Avenue

This imposing brick mansion with a Greek Revival style portico was built about 1810 by George P. Macculloch. That enterprising gentleman of Scottish origin conceived the Morris Canal in 1823 and founded the Morris Canal and Banking Company to finance construction of the waterway. The Canal connected the Delaware and Hudson Rivers and was a highly successful enterprise until the advent of the railroads in the 1860s. (See Stanhope-Waterloo Village.)

In recent times a philanthropist W. Parsons Todd has generously redone this mansion to illustrate how a well-to-do gentleman like George Macculloch would have lived in the mid-nineteenth century. The interior has been filled with a wealth of antiques, china, crystal, silver, and paintings. For instance, as you enter the front hall you will see a portrait of George Washington painted by the famous Charles Willson Peale. Fanciers of antique furniture will enjoy a fine array of English antiques including chairs from many periods. There are framed descriptions of the more noteworthy pieces.

Macculloch Hall was once the home of George P. Macculloch, who was responsible for the Morris Canal. Thanks to the generosity of W. Parsons Todd, the house has been restored and furnished with quality antiques. (Courtesy Macculloch Hall)

This is the elegant dining room in Macculloch Hall. (Courtesy Macculloch Hall)

Of the rooms on the main floor, the dining room is particularly appealing, sparkling with its silverware and crystal. The table is set with a rare collection of Dr. Wall Worcester china and is so inviting that one wishes to be asked to partake of a meal in this elegant setting.

Upstairs there are several bedrooms furnished with English and American furniture. There is also an exhibit of Thomas Nast cartoons. Nast was the noted caricaturist of the late nineteenth century who penned the elephant as the Republican Party's symbol and the donkey as that of the Democratic Party. He lived across the street from Macculloch Hall.

DRIVING DIRECTIONS: From the Morristown Green, head south on Route 202 a short distance to your first left, onto Macculloch Avenue. The house is in the third block and on your right.

Open: By appointment only
Admission: Free
        (R)

The tour is semi-guided and lasts about 30 minutes. There is an eight-page pamphlet on the house, its history and contents, available at no charge.

Schuyler-Hamilton House (c. 1765)
5 Olyphant Place

This two-story Colonial dwelling is well known locally, as it was here that Alexander Hamilton courted and became engaged to Betsy Schuyler in the winter of 1779–1780.

Downstairs, one visits the living, dining, and east rooms, which are furnished with period antiques. Upstairs in the bedroom, a lock of Betsy Schuyler-Hamilton's hair together with her widow's cap and collar are on exhibit.

In Morristown one can visit the Schuyler-Hamilton House, which is owned by the Daughters of the American Revolution. It is here that Alexander Hamilton courted his future wife, Betsy Schuyler. (Author's own photograph)

DRIVING DIRECTIONS: From the center of Morristown head east on Morris Street about a half mile, then take the second left under the railroad bridge onto Olyphant Place. The house is the first one on your right.

Open: Tuesday and Saturday 2 P.M.–5 P.M. or by appointment.

Admission: Free

The guided tour lasts about 15 minutes. One can purchase a four-page description of the house for 10¢. Postcards, notepaper of the house, and dolls made by the DAR (who own the house) are for sale.

Speedwell Village
333 Speedwell Road

In recent years a group of historic preservationists saved the Speedwell factory and barn and the George Vail homestead from falling into a state of complete disrepair.

Two important developments took place here. The Speedwell Iron works (which the Vail family ran from 1814 to 1873) produced and installed the engine and some other parts for the *S.S. Savannah,* the first ship to cross the Atlantic under steam power. Later, in 1838, the first telegraph message was transmitted from the Speedwell factory by Samuel F. B. Morse and Alfred Vail. Young Vail had met Morse in New York City and he and the Vail family

This was the Vail Homestead and is located near the barn-factory at Speedwell. (Author's own photograph)

From this barn-laboratory the first words were transmitted by telegraph on January 8, 1838. The inventor, Samuel F. B. Morse, had been helped by Alfred Vail, whose family ran the Speedwell Iron Works at this location. (Author's own photograph)

provided the inventor with scientific and financial assistance and made the Speedwell laboratory available to him. It was from that laboratory on January 8, 1838, that the first words "a patient waiter is no loser" were sent on the now famous electromagnetic telegraph.

The three buildings mentioned above have now been renovated on the outside. The red barn and cottage were apparently built about 1808. The factory (which was also a grist mill at one time) is of undetermined age, but it appears that the second floor where Morse and Vail conducted their experiments in the 1830s has not been altered since that time.

Several other historic houses have been brought into Speedwell Village. These include the Ford Cottage, probably dating from the early 1800s; Lhomedieu House, portions of which are believed to predate the revolution; and Este House, which dates to about 1786. To the present, the Speedwell group has concentrated on preparing the grounds and repairing the buildings. They will now proceed with historical and architectural surveys of the buildings. For that reason, their dates can only be estimated now.

In time this active group envisions furnishing the homestead and establishing exhibits and demonstrations on the Speedwell Iron Works, the Vail and Morse families, and the telegraph. We wish them well in their worthwhile endeavors.

DRIVING DIRECTIONS: Proceed northwest on Route 202 (Speedwell Road) from the Morristown Green about one and a quarter miles, where you will see the Homestead on your right.

Open: By appointment only. The Village is not yet officially open so that it is too soon to determine the times of opening, duration of the tour, and admission.

Morristown National Historical Park

It was in Morristown and its environs that General Washington and his troops spent the long cold winter of 1779–1780 during the Revolutionary War. Washington, accompanied by his wife, Martha, stayed in comparative splendor at the Ford Mansion in town, while his troops built wooden huts about five miles away in Jockey Hollow. Some officers found accommodations in nearby farmhouses, like Wick House.

It is indeed fortunate that both of these houses are still in good condition and that the National Park Service has

In this section of the Morristown National Historical Park one can visit Ford Mansion, where George Washington stayed during the winter of 1779–1780. (Courtesy National Park Service, Morristown National Historical Park, Morristown, New Jersey)

done a splendid job in recreating the full atmosphere of this historic winter encampment. It is advised that you start at the Ford Mansion, which is easy to find, and there you can pick up a free four-page guide and map to the various sections of the Park. A visit here is recommended.

Ford Mansion (c. 1774)
230 Morris Street

Colonel Jacob Ford, a wealthy iron producer, had this

This is the Colonial kitchen in the Ford Mansion. (Courtesy National Park Service, Morristown National Historical Park, Morristown, New Jersey)

mansion built about 1774, using architectural designs recently received from England. The Palladian effect, then still in vogue in England, is seen in the treatment of the archway over the front door, which is flanked by two smaller rectangular windows. This pleasing and symmetrical design is carried out in the second story windows above the front entrance.

Fortunately, the mansion has been preserved, having been owned by the Ford Family until it was sold to the Washington Association in 1873. The National Park Service took over its operation in 1933.

The Ford family generously left some of the original furniture that was there during Washington's stay. This obviously makes the tour all the more exciting and worthwhile. Some of these antiques seen are a Chippendale-style secretary desk that Washington used (in the living room) ; a finely carved Chippendale dressing table, highboy, and one of his large campaign chests (in his and Mrs. Washington's bedroom) ; and some Windsor chairs (in Washington's office). The rest of the well-appointed furnishings are in keeping with the period of the house and are primarily of the Queen Anne and Chippendale styles. However, an interesting piece in one of the upstairs bedrooms is a rare folding bed dating from the Revolutionary War period.

DRIVING DIRECTIONS: From the Morristown Green take Morris Street east about one mile. You can't miss the house, which will be on your left.

Open: Daily 10 A.M.–5 P.M.
Closed only on Thanksgiving, Christmas, and New Year's Day

Admission: Adults—50¢
(R) Children under 16—free when accompanied by an adult

The tour is self-guided, aided by descriptive placards posted outside each of the 11 rooms, and lasts about 45 minutes. There are books, pamphlets, postcards, and souvenirs for sale at the museum.

Note: The price of admission also includes entrance to the Historical Museum, behind the Ford Mansion. This houses an outstanding collection of manuscripts on Revolutionary times, collections of Revolutionary arms, objects that have been dug up at Jockey Hollow, and exhibits of Early American pottery, silver, pewter, and glass. Among the Washington memorabilia viewed is the silk suit that he wore when inaugurated as President, April 30, 1789.

Wick House (c. 1750)
Morristown National Park

Henry Wick came here from Long Island in about 1748 and took up farming. His shingled and weatherboarded farmhouse was completed in about 1750 and reflects the influence of New England architecture, having one chimney, low eaves, and a high-pitched roof line. Some of Washington's officers made it their quarters during the winter of 1779–1780.

The Wick House is well known for the legend that Temperance Wick, daughter of the owner, hid her horse

In another section of the Morristown National Historical Park one can visit Wick House, where some of Washington's officers were quartered in 1779–1780. (Courtesy National Park Service, Morristown National Historical Park, Morristown, New Jersey)

Wick House has an attractive kitchen. The featured table can also be used as a chair and storage box. (Courtesy National Park Service, Morristown National Historical Park, Morristown, New Jersey)

in her bedroom on January 1, 1781, thereby saving the horse from the mutinous Pennsylvania brigade, who were revolting for lack of pay and food.

Afterwards it continued to be a farmhouse and, amazingly, was never modernized. The National Park Service acquired it in the mid-1920s. This is a good example of a Colonial farmhouse and only some of the panelling and floor boards have had to be restored.

The interior is tastefully furnished with Colonial pieces and appointments. An interesting item is the kitchen table, which can also be used as a chair and as a storage box. In the children's bedroom is an unusual double children's

Behind Wick House is Jockey Hollow, which has numerous reconstructed huts like this where Washington's troops spent the cold winter of 1779–1780. (Courtesy National Park Service, Morristown National Historical Park, Morristown, New Jersey)

wagon seat. In the pantry observe the wooden yoke, which was put over a goose's head in order to prevent its going through a fence. An herb and flower garden is outside.

DRIVING DIRECTIONS: From the center of Morristown, take Washington Street (Route 24) west, three blocks to Western Avenue, where you turn left. Continue for about four miles on this road, which becomes Jockey Hollow Road (this road has numerous markers describing historic events that took place here). The road ends

at Tempe Wick Road, where you turn right, and the
house is on your right.

Open: Daily 1 P.M.–5 P.M.
              Closed December and January

Admission: Free

The self-guided tour is aided by posted placards describing the contents of the five rooms. After leaving, be sure to follow the signs to see the reconstructed cabins and hospital of Washington's encampment in Jockey Hollow. The park also has picnicking tables, but no fires are allowed. Hiking enthusiasts will be delighted to know that there are miles of trails. Ask a member of the park staff for a map.

MOUNT HOLLY

Mount Holly is the county seat of Burlington County, and has some very attractive Colonial and Federal buildings. These are described in the pamphlet, *A Walking Tour of Mount Holly,* which may be obtained free of charge at the Township Hall at 23 Washington Street or at the more centrally located Prison Museum. Notable among these buildings are the Colonial Court House, built in 1796, and the Friends Meeting House, erected in 1775.

Burlington County Prison Museum (c. 1808)
128 High Street

The prison was designed by Robert Mills, the Federal architect, in 1808, and became the model for other penal buildings in this country. It was in use as a prison until 1965. Unlike the other buildings in this guide, which are included for their architecture, furnishings, or historic qualities, the prison is presented as a granite document of rooms that are not filled with antiques but which do present another area of this country's social development.

The Burlington County Historical Association has maintained the building virtually as it was, and there are limited permanent exhibitions of the inmates' handicrafts.

In Mount Holly a unique "historical house" is the Burlington County Prison whose architect was the well-known Robert Mills. No longer used as a prison, it now contains some exhibits pertaining to penal life. (Courtesy the Burlington County Prison Museum)

The Association plans to enlarge the exhibits of prison life and hopes to include other features about the history of the county. Presently, they do have changing art exhibits. They even propose to present summer concerts in the former residence yard. We wish them well in the execution of these plans.

DRIVING DIRECTIONS: Take Exit 5 on the New Jersey Turnpike. Then go east on Route 541 for about two and a half miles. Route 541 is High Street. The prison is #128 on your right.

Open: Tuesday–Saturday 10 A.M.–12 P.M., 1 P.M. –4 P.M.

Admission: Free

   (R)

Also in Mount Holly is this one-room schoolhouse which was reconstructed and furnished by the National Society of Colonial Dames in the State of New Jersey.

The tour is guided and takes about 20 minutes. Plates, postcards, and notepaper showing the prison are for sale. A four-page brochure describing the prison is available free of charge.

John Woolman Memorial (c. 1783)
99 Branch Street

Located on property that belonged to the Quaker Abolishionist, John Woolman (1722–1772), this three-story brick house was built in 1783 on the occasion of his daughter's marriage. The front porch was added in the early nineteenth century, and it and the house need some refurbishing.

The building is now a Quaker center and retreat, and it contains a collection of books by and about Woolman and some period furnishings such as his rocking chair in the parlor. The bedrooms on the upper floor feature old quilts.

DRIVING DIRECTIONS: Same as for the prison to High Street, driving past it and the Court House to the traffic light at Garden Street. (The Friends Meeting House is on the corner.) Turn left on Garden Street about seven-tenths of a mile to Branch Street. Turn right, and the house is on your left, #99.

The John Woolman Memorial in Mount Holly is now a Quaker retreat and also contains books and memorabilia pertaining to the Quaker Abolishionist. (Author's own photograph)

Open: Monday–Saturday 9 A.M.–6 P.M.
            Sunday 1 P.M.–6 P.M.
Admission: Free
      (R)

There is a guided tour of about 15 minutes. A two-page brochure of the house plus a map are available free of charge. Picnicking facilities are available on the grounds.

The Old Schoolhouse (c. 1759)
35 Brainerd Street

John Woolman may have taught in this building, which is constructed in the Flemish bond style. The one-room school house has been restored and the bricks on the front and sides are original.

DRIVING DIRECTIONS: From High Street take your first left after the traffic light at the Friends Meeting house (Garden Street). You are now on Brainerd, and the school is a block on your left.

Open: Wednesday 1 P.M.–5 P.M., May–October.
            Also by appointment
Admission: Free

Since this is a one-room schoolhouse, it may be viewed by peering through the windows.

NEWARK

New Jersey Historical Society
230 Broadway

Although this building does not truly qualify as a historic house, it is the headquarters of the New Jersey Historical Society and contains a wealth of information and material on all aspects of New Jersey's history and culture. To highlight a few of these, there are several exhibits of antique silver, porcelain, and china, some being New Jersey or American made and others European. Be sure to take note of the silver teapot made by Elias Boudinot about 1750. His son, Elias Boudinot, Jr., was a leading advocate for the Colonists' cause and later President of the Continental Congress (see Boudinot House: Elizabeth).

The Society saved the Antill-Ross parlor from demolition and moved and installed it here. This parlor was part of a house located near New Brunswick, built about 1739. The fireplace is flanked by two well-proportioned built-in cupboards with hand-carved shell motifs. The panelling is most handsome as is the room itself, which is furnished with Queen Anne and Chippendale antiques.

Life in New Jersey in 1875 is portrayed in three small

The headquarters of the New Jersey Historical Society is in this building in Newark. It contains an abundance of information on all aspects of New Jersey. (Courtesy New Jersey Historical Society)

facsimile rooms: a bedroom, parlor, and kitchen, all fitted with Victorian furnishings. Another room contains Hepplewhite antiques.

The New Jersey Historical Gallery provides a unique and interesting presentation of the state from earliest to modern times. This is shown with dioramas, scale models, photographs, maps, old manuscripts, and displays of pertinent items such as Early American cooking utensils and weapons, and Indian relics and utensils.

A visit here is recommended. The Society is presently expanding its exhibits. Incidentally, for want of space the

This picture shows a corner of the Ross-Antill Parlor at
the New Jersey Historical Society's headquarters. The
panelling was moved here from the Ross-Antill House in
New Brunswick. The two Queen Anne pieces are of ex-
ceptional quality. (Courtesy New Jersey Historical Society)

Society has generously loaned some of its collections of antiques to several historic houses in the state, most of which
are described in this book.

DRIVING DIRECTIONS: I always seem to get lost in Newark and have found that the easiest way to get to the Society is to get off the Garden State Parkway at Exit 148. Follow Bloomfield Avenue east nearly two and three quarter miles to Fourth Avenue. (This is the street immediately after one of the main thoroughfares, Mount Prospect Avenue.) Take a left on Fourth Avenue and continue a short distance until you reach Broadway. The Society is in the second block north of there.

Open: Tuesday—Saturday 10 A.M.—4:30 P.M.
               Closed Sunday, Monday, and the month of
               August
Admission: Free
        (R)

The tour is not guided and lasts about one hour. Most items are described on placards. Guided tours can be arranged in advance. Books that the Society publishes on all phases of New Jersey history, notepaper, and souvenirs are for sale.

Note: The Society has an outstanding library covering New Jersey history. A transportation museum, with emphasis on railroads and waterways, is located on the third floor.

NEW BRUNSWICK

Buccleuch Mansion (c. 1734)
Buccleuch Park

This is a handsome three-story mansion in the Georgian style with noteworthy fanlight on the top floor. Of particular interest inside is the most unusual and priceless French wallpaper in the wide hall downstairs and in the smaller hall upstairs. This was hand painted in France by DuFour and hung here in 1819. The colors of the paper have faded and some sections are peeling but it remains very attractive.

The rooms are furnished in different periods. The best appointed one, as well as the earliest period represented, is the Queen Anne drawing room, which has been restored. The furniture in it, particularly the sofa, chairs, and lowboy, is of good quality and dates from 1740 to 1750. Also take note of the grandfather's clock, made in 1762 by a local clocksmith, Peter Luepp. The case was crafted by Matthew Egerton, a New Brunswick cabinetmaker. In contrast to this earlier room, the later Federal dining room displays some locally-built furniture, including a Sheraton-style sideboard made by Oliver Parsell in 1811. The later Victorian parlor contains a good selection of the ornate furniture so popular then.

Buccleuch Mansion in New Brunswick has an attractive fanlight. It is owned by the City of New Brunswick and is well maintained by the Daughters of the American Revolution. (Author's own photograph)

The bedrooms on the second and third floors are furnished in periods ranging from Federal through Victorian. The third floor also has a room with toys and children's furniture of the 1880s. Another room on that floor has displays of colonial crafts.

DRIVING DIRECTIONS: Take Exit 9 on the New Jersey Turnpike. Then follow Route 18 (Kennedy Memorial Parkway) about two miles west until it deadends at Route 27. You go left (or south) on Route 27 for a few blocks to George Street. Turn right on George Street and follow it about three quarters of a mile. You will see the man-

sion perched on a hill in Buccleuch Park, which will be
on your left. (Along George Street you will be driving
through a section of the New Brunswick campus of
Rutgers University, the eighth oldest college in the
country, founded in 1766.)

Open: Saturday and Sunday 3 P.M.–5 P.M., the end
             of May through the last weekend in Octo-
             ber, or by appointment

Admission: Free

        (R)

The tour is not guided and lasts about 30 minutes.
There is a free brochure (four pages) describing the
history of the house and its contents. Picnicking facilities
are available in the park, which also has a playfield.

OCEAN CITY

Ocean City Historical Museum (c. 1913)
409 Wesley Avenue

Although this red brick institutional building does not qualify as a historic house, it is included here because it contains a Victorian section. As you wander through the four rooms (parlor, dining room, kitchen, and bedroom), you will feel like you are back in the 1890s, when this seaside resort began to flourish.

Other exhibits in this old school are devoted to Ocean City's history, Indian artifacts, old dresses and costumes, and the *Sindia,* a cargo ship that was wrecked on Ocean City's shores in 1901.

DRIVING DIRECTIONS: Take Exit 30 on the Garden State Parkway. You will be on Laurel Avenue, which becomes Route 52. Follow this route about three miles, around the traffic circle and over the bridge into Ocean City. After you enter Ocean City, Route 52 becomes 9th Street. Continue on 9th Street about five blocks and then turn left on Wesley Avenue. The museum will be about one-half mile on your right.

This is not a historic house but is included since the Ocean City Historical Museum is located here. (Courtesy David N. Poinsett)

Open: Monday–Saturday 10 A.M.–4 P.M., Summer
           Tuesday–Saturday 1 P.M.–4 P.M., rest of
           the year
Admission: Free
    (R)

This Victorian room in the Ocean City Museum depicts life in that town in the 1890s. (Courtesy David N. Poinsett)

The tour is semi-guided and lasts about 20 minutes. There is a four-page brochure describing the museum available at no charge. Postcards, stationery, and some gift items are for sale.

Take advantage of the lovely boardwalk and beaches in Ocean City.

PATERSON

Lambert Castle (c. 1892)
5 Valley Road, in Garret Mountain Reservation

A boyhood dream was realized when Catholina Lambert built this castle in 1892. An immigrant from England in 1851, Lambert made his fortune in the manufacture of silk and ribbon. This impressive structure is constructed of local brown and grey stone. No longer furnished as a residence, the castle's noteworthy features are the carved mantelpieces and decoratively painted plaster ceilings.

Five rooms are open to the public, as is the hall, which has lovely oak panelling, a seventeenth-century Italian painting, and a Tiffany lighting fixture in the ceiling. As you look up the oak staircase, note the handsomely carved oak picture frame at the bottom, which is unusual since it is empty. It is said that Lambert romantically had this frame put there so that he could gaze upon his wife as she came down the stairs as though she was in a picture. The library, which now houses the Passaic County Historical Museum's library, has its original gold-leaf ceiling.

The McKinley-Hobart Room has a lavishly carved mantel and chandelier and features displays on Garrett A. Hobart, a native of the area, who was Vice President

One can see the skyscrapers in New York from Lambert Castle in Paterson. Built by a wealthy industrialist, it is now maintained by the Passaic County Historical Society and the Passaic County Park Commission. (Courtesy Passaic County Police)

during McKinley's first term. (Hobart died in 1899. Had he lived and been re-elected with McKinley, he would have assumed the Presidency upon McKinley's death by assassination.)

One room, rich with its painted ceiling and walls of pressed plaster, contains a collection of over 2,000 silver spoons.

Note: The view from the castle can be terrific. On a clear day one can see the Verrazano Bridge, Manhattan's skyscrapers, and the George Washington Bridge. The more ambitious may wish to ascend

the 70-foot tower behind the castle for a more
commanding view. Mr. Lambert built this tower as
his observatory.

DRIVING DIRECTIONS: Take Exit 155 on the Garden State
Parkway. Proceed straight ahead for one block, then
turn right on Valley Road. You will almost immediately
see the castle high on the hill to your left.

Open: Wednesday, Thursday, Friday 1 P.M.–5 P.M.
                Saturday and Sunday 1 P.M.–4 P.M.

Admission: Free

    (R)

There is no guide but all displays are well labeled and
a visit lasts about 20 minutes. Books and pamphlets on
New Jersey can be purchased. A bulletin on the castle and
its contents, which was prepared by the Passaic County
Historical Society, costs 25¢. Slides of the castle are also for
sale.

One may picnic and charcoal and walk in the woodland
park surrounding the castle. In the summer there are band
concerts in the park. For information regarding these and
other activities in the park, write to the Passaic County
Park Commission, Lambert Castle, Paterson, New Jersey.

PLAINFIELD

Drake House (c. 1745)
602 West Front Street

This is the Nathaniel Drake House, built in 1745 by a prosperous farmer, Isaac Drake, for his son, Nathaniel. General George Washington spent time at this farm house at the time of the battle for the Watchungs, also known as The Battle of the Short Hills (June 25–27, 1777). Among items connected with him and the battle are the pine tea table in the dining room, on which it is said that Washington did some writing, and a diorama in the basement depicting the battle for the Watchungs.

Originally the house consisted of a kitchen, dining room, living room, and bedroom downstairs with additional sleeping quarters upstairs. In 1860 the house was expanded and redone in the Victorian style as may be seen in the turret, tower, and "gingerbread" effect around the dormer windows. The house is, therefore, of mixed architectural parentage but, to me, pleasing to the eye.

Similarly, many periods of furnishings are seen inside: a Colonial kitchen with its original fireplace and ovens. Colonial dining and bedrooms, and an Empire living room. Also note the colorful stencilling on the floor of the

148

During the battle for the Watchungs (June 25-27, 1777),
General Washington stayed at this farmhouse in Plainfield.
The Victorian remodeling took place in about 1860. (Cour-
tesy Historical Society of Plainfield and North Plainfield)

living room and on the wall of the bedroom. This is
original to the house and was often found in Colonial
dwellings. Rotating exhibits are planned. Be sure to see
the Jacobean-style oak chair that was fashioned of planks
taken from a British ship that was sunk in the Delaware
River during the Battle of Red Bank, October 22–23, 1777.

The house is owned by the city of Plainfield and is well
maintained and operated by the Historical Society of
Plainfield and North Plainfield.

DRIVING DIRECTIONS: Plainfield is located south of Route 22.
Leave Route 22 at the exit for Plainfield and proceed
south on Route 531 (Watchung Avenue, which bears
right onto Somerset Street) for about one and a half
miles where you turn right onto West Front Street

Another room to visit at Drake House is this Empire-style parlor. Drake House is owned by the city of Plainfield and is maintained and operated by the Historical Society of Plainfield and North Plainfield. (Courtesy Historical Society of Plainfield and North Plainfield)

Among the rooms one visits in Drake House is this Queen Anne dining room, which is furnished as in the period of George Washington. (Courtesy Historical Society of Plainfield and North Plainfield)

(Route 28). The house is a short distance on your right.

Open: Monday, Wednesday, and Saturday 2–5 P.M.
             or by appointment

Admission: Free

     (R)

The tour is self-guided, lasting about 20 minutes. Good descriptions of each room are posted. A guided tour may be arranged upon request in advance. A four-page brochure describing the house is available free. Postcards, slides, and notepaper of the house are for sale.

Picnicking facilities are available in Greenbrook Park. Drive past Drake House on Front Street about one mile and turn right on West End Avenue. The entrance to the park is one-half block on the left.

PRINCETON

Known primarily for its university, Princeton is architecturally and aesthetically one of the most pleasing towns in New Jersey. The town was founded in the late seventeenth century as a farm community, and assumed its present name in 1724. In later years its importance was as a stage stop between Philadelphia and New York. In Colonial times it was a principal political center. The Battle of Princeton on January 3, 1777, was an important victory for the colonists' cause. The Battlefield Park is located one and a half miles from the center of town on Mercer Road. In 1783 the Continental Congress met at Nassau Hall and Princeton was in effect the capital of the country. At that time Washington stayed in Rockingham and was a frequent guest at Morven.

Due to the semi-secluded character of the town, most of the buildings with historical significance remain, and the town has examples of American architectural styles from its earliest days. Those that are open to the public are described here. Others, together with the University and church buildings, are listed in the walking tour of Princeton provided in the *Illustrated Historic Fact Book on Princeton,* which is available for $1.00 at Bainbridge House and at local stationers and book stores. This also contains a driving tour of Princeton. Another valuable map and guide for motorists is *Follow in Washington's Footsteps,* which enables you to trace the local events of the Revolutionary War from Washington crossing the Delaware in 1776 onwards. This afternoon drive also allows you to see some lovely, unspoiled countryside. This pamphlet can be purchased for 25¢ at Bainbridge House.

Also take advantage of the free, hour-long tour of the Princeton University campus. These tours depart four times daily (9:40 A.M., 11:40 A.M., 1:15 P.M., and 3:20 P.M.) and twice on Sunday (1:15 P.M. and 3:20 P.M.) from Stanhope Hall, which is to the right of Nassau Hall. It is advisable to write (or call: 609–452–3603) ahead for a reservation (Orange Key Guide Service). A trip to the town and visits to the houses open to the public are recommended.

Bainbridge House (c. 1765)
158 Nassau Street

This Georgian brick and clapboard town house was built about 1765 by Robert Stockton, a member of the well-known Princeton family. When the Continental Congress convened in Nassau Hall in the fall of 1783, this house was one of the places listed to accommodate the delegates. The house is named for William Bainbridge, who was born here in 1774 and became the famous Naval Commodore. He is best remembered for commanding the *Constitution,* nicknamed "Old Ironsides," during the War of 1812. His father, Dr. Absalom Bainbridge, rented this house from Robert Stockton and had his medical office here.

Another physician, Dr. Ebenezer Stockton, resided and practiced medicine here from 1799 until 1835. He was responsible for redoing the parlor about 1815.

Today there are three rooms to be seen, containing

some pleasing antiques such as the English tall-case clock made by Thomas Wagstaff of London in about 1790, the Queen Anne-style chest on a frame made in Delaware about 1750, and a Sheraton-style settee built about 1800. It is interesting to contrast the original built-in corner fireplace with panelling in the Doctor's office (the fireplace and panelling were moved from upstairs) with the "newer" fireplace in the early nineteenth-century parlor. The later fireplace has a mantel and is adorned with classical motifs.

The Historical Society of Princeton, which leases this house from Princeton University, is in the midst of an am-

The headquarters of the Historical Society of Princeton, Bainbridge House, is named after the well-known Naval Commodore, William Bainbridge, who was born here. (Courtesy David N. Poinsett)

bitious program to expand its collections and has restored the exterior of the house. We wish them well in their worthwhile endeavors.

DRIVING DIRECTIONS: Take Exit 8 on the New Jersey Turnpike, and then drive west on Route 33 one mile to the traffic light in the center of Hightstown. Drive west on Route 571 seven miles until it ends at a traffic light in Princeton, which is at Nassau Street. Take a left and Bainbridge House is the first house on your right (next to the movie theater).

Open: Monday–Friday 10 A.M.–3 P.M.
              Saturday 1 P.M.–3 P.M.
              Sunday 2 P.M.–4 P.M. or by appointment
              During June, July, and August the weekly hours are 10 A.M.–1 P.M. (The Society is trying to recruit volunteers so that the summer hours can be extended. Be sure to check the hours if you are in the vicinity.)

Admission: Free
       (R)

The tour is self-guided and lasts about fifteen minutes. There are descriptive placards in each room. Postcards and notepaper of the house, maps of New Jersey, and books on Princeton and its environs are for sale. The *Illustrated Historic Fact Book of Princeton* and *Follow in Washington's Footsteps,* previously mentioned, are available here (the former costing $1.00 and the latter 25¢).

Note: Changing exhibits are displayed in the newer addition at the back of Bainbridge House. There is also an extensive library on state and local history and genealogy.

Drumthwacket (c. 1836 with later addition c. 1899)
Stockton Street

In about 1836 this white house with handsome columns was built in the Greek Revival style. The side wings were

This magnificent mansion, known as Drumthwacket, is located in Princeton. Presently owned by the state, its future use is undecided. (Courtesy David N. Poinsett)

added on in the late 1890s. Drumthwacket was purchased by the State in 1966. Since then it has been painted outside. There had been some discussion of making it the Governor's residence as Morven is too small, but this was decided against. A decision regarding the future of Drumthwacket is awaited. We hope that whatever the outcome, this elegant mansion will be furnished and made open to the public as is Morven now, and thus we include it in this book.

If you are in Princeton, you may wish to drive down and look at Drumthwacket, whose exterior with tall Ionic columns is particularly handsome. The grounds are also very attractive.

DRIVING DIRECTIONS: This house is on the same street as
Morven (Stockton Street), about one and a half miles
south on your left. See driving directions to Morven.

Morven (c. 1701, with later additions of unknown dates)
55 Stockton Street

When this early Georgian-style house was built in 1701
as the Stockton Manor House, it certainly must have been
the talk of the area, as it was much grander than the com-
paratively rudimentary earlier Colonial dwellings in the
town. The classical columns and detailed pediment of the
porch illustrate the influence of the Italian Renaissance in

This is Morven, the official residence of the Governor of
New Jersey in Princeton. It can be visited by appointment.
(Courtesy N.J.D.O.E.P.)

This is an interior view of the Governor's residence, Morven. (Courtesy David N. Poinsett)

architecture, which was also in vogue in Europe at that time.

The unmatching wings were doubtlessly added at separate times, although it is difficult to pinpoint because there are no records. The house has also suffered two fires, and it may be that they caused this reconstruction.

Morven is associated with numerous events: one owner, Richard Stockton (grandson of the builder), signed the Declaration of Independence; the British General Cornwallis made his headquarters here in late 1776 and early 1777 during the Revolutionary War; and while the Continental Congress met in Princeton in 1783, George Washington (who was residing in nearby Rocky Hill) visited here.

How fitting it is that a house with so many historical ties is the present official residence of the governor of New Jersey. In addition to the governor's personal furnishings, there are some antiques and reproductions owned by the State. In the dining room there is a Regency antique dining room table, around which are reproduction Chippendale chairs. The permanent collection also includes some portraits of the Stockton family and their contemporaries, a set of French chairs with cane seats sent to the Stockton family by a French diplomat after the Revolutionary War, and a handsome English grandfather's clock made by George Guelst in about 1780. Four rooms and two halls are open to the public.

DRIVING DIRECTIONS: Follow directions to Bainbridge House. On your left will be Princeton University with Nassau Hall. If you continue on Nassau Street about one mile (or two traffic lights), Nassau Street runs into Stockton Street (also Route 206). Morven is on your right shortly after the intersection.

Open: Tuesday 2 P.M.–4 P.M., January–June
               September–November by appointment
               only. Write to the Governor's Social Secre-
               tary, Mrs. Eleanor Wright, c/o Morven.
Admission: Free
          (R)

The tour is semi-guided and lasts about 15 minutes. Two four-page brochures, one describing the history and architecture of Morven and the other describing the furnishings, are available free.

Rockingham (c. 1734, later addition 1764)
Route 518, Rocky Hill

George Washington and his wife, Martha, were indeed fortunate that this gracious house was leased for them while the Continental Congress met in Princeton at Nassau

When the Continental Congress met in Princeton in 1783, George Washington and his wife stayed here at Rockingham. From the second floor porch, General Washington gave his "Farewell Orders to the Armies." (Courtesy N.J.-D.O.E.P.)

Hall from late August until mid-November, 1783. Here in the upstairs study Washington wrote his "Farewell Orders to the Armies." He subsequently delivered this famous speech from the second-story porch to a small cadre of troops, his personal bodyguard, gathered below.

The history of Rockingham dates to 1734, when a small two-story farmhouse was built with one room on each floor. The house was expanded in 1764 and the two-story porch added on after John Berrien purchased this property. This two and a half story white clapboard house is therefore referred to as The Berrien Mansion as well as Rockingham.

The interior of the house possesses many of its early features such as the fireplace and handsome pine panelled wall in the dining room. The dining room and the room above it were the original two rooms of the house. Note in each corner of this room and the one above (George Washington's study) the unusual beams that structurally support the house and are known as "gun stalk uprights." In the living room there are hand-carved cornices of the period.

Rockingham is tastefully furnished throughout with eighteenth-century antiques of the period when the Washingtons lived here. There is a Hepplewhite sideboard in

It is in this room adjacent to the balcony of Rockingham, that Washington apparently wrote his Farewell Orders. (Courtesy N.J.D.O.E.P.)

The dining room at Rockingham near Princeton is attractive. (Courtesy N.J.D.O.E.P.)

the dining room, and in Washington's bedroom there is a Chippendale combined blanket and chest of drawers.

Another appealing feature about Rockingham is that there are numerous unusual items to be seen. Among these are the "oil-time" lamp and "wag-on-wall" clock in Washington's study. (The latter refers to the movements of a long-cased clock before the case has been added.) In Washington's dressing room, observe the shaving bowl, shaving mirror, and well-disguised potty-chair or commode. In the dining room, a picture on the wall exhibits what is called "papyrotamia," an old craft whereby paper is cut into miniscule designs or scenes, then pasted on a backing, glazed, and framed. One must see this to appreciate the fine precision work this art form required.

A visit to Rockingham is recommended in order to see George Washington's headquarters and to visit a well-appointed Georgian residence.

DRIVING DIRECTIONS: From the center of Princeton (Nassau Street, Vandeventer Avenue, and Washington Road—near Bainbridge House) drive north on Route 27 (Nassau Street) three miles to the center of Kingston, where you take a left on Laurel Avenue and proceed three quarters of a mile until you deadend into Route 518. Take a right and drive half a mile until you see the house on your left. There are several signs directing you to Rockingham from Kingston. As you approach the house, don't be concerned because you don't see the two-story porch, for you are coming into the back of the house, and the porch is across the front.

Open: Tuesday—Saturday 10 A.M.—12 P.M., 1 P.M. —5 P.M.

Sunday 2 P.M.—5 P.M.

Closed Monday, Thanksgiving, Christmas, and New Year's Day

Admission: Adults—25¢

(R) Children under 12—free

The tour is guided and lasts about 40 minutes. There are also posted descriptions of the contents of each of the ten rooms viewed. There is a four-page brochure available at no charge. Postcards and notepaper of the House, together with souvenirs of George Washington's stay, are for sale.

RAMSEY

Old Stone House (c. 1743)
536 Island Road

In 1955 the Old Stone House, a quaint Dutch-style house with a gambrel roof, was threatened with destruction by the State in order to make room for an overpass on Route 17. The Ramsey Historical Association and the borough saved it from this unhappy ending. Since taking over the house at that time the Association has been restoring it.

Two rooms are visited, one of which has a restored fireplace with panelling around it. The fireplace is adorned by an array of Colonial kitchen utensils and equipment. The original exposed beams with bead molding are structurally interesting.

The well-appointed Colonial antiques and accessories illustrate the surroundings of an average Early American household. Of particular note is a massive Dutch kas (cupboard) in the west room. Currently the Association is preparing a third room, a bedroom in the Colonial style.

It is thought that at one time the Old Stone House was a tavern, probably during the period of the Revolutionary War.

The charming Old Stone House is located in Ramsey in northern New Jersey. It is maintained by the Ramsey Historical Association. (Author's own photograph)

DRIVING DIRECTIONS: Driving north on Route 17, do not exit at the first exit that says Ramsey. Get off further north at the Spring Street Exit, then loop over Route 17 on the overpass and the house will be on your right immediately after you leave the bridge. Heading south on Route 17, you exit at Spring Street, immediately after leaving Route 17 you will see the parking lot of the house on your left.

Open: Sundays 2 P.M.–4:30 P.M. May through October or by appointment

Admission: 25¢

(R)

The tour is guided and lasts about 15 minutes. A booklet on the history of the house is available for 25¢. Postcards and notepaper are also for sale. Picnicking, fireplaces, and nature trails (with labeled trees) are to be found in the Campgaw Mountain Reservation on Route 202 in nearby Mahwah.

The building behind the Old Stone House contains exhibits of old toys, games, clothes, and books.

Note: Railroad buffs will be interested in the *Old Station Museum* in nearby Mahwah. The Mahwah Historical Society has taken over this Erie R.R. station that dates to 1874. The Museum contains exhibits of old railroad equipment. You can also see an old Erie R.R. caboose on Old Station Lane. It is open to the public on Sunday afternoons during the months of April through December.

RIDGEWOOD

Paramus Historical and Preservation School House Museum
(c. 1872)
650 East Glen Avenue

This large one-room schoolhouse in the Victorian style was originally affiliated with the adjacent Old Paramus Reformed Church. In the 1870s the church school became part of the Ridgewood system and was used until 1905.

Vestiges of Victorian school days are seen in the original desks, potbelly stove, lamps, and restroom with its three-hole seat. The "good old days" are also remembered in the Victorian parlor, with its period furniture, glassware, and china, and in the Lincoln kitchen, equipped with utensils and gadgets.

Since this area of New Jersey was largely settled by the Dutch, there is also included a facsimile late eighteenth-century Dutch kitchen.

There are numerous other exhibits including Indian artifacts, toys, dolls, and artifacts of the Old Reformed Church.

DRIVING DIRECTIONS: Heading south on Route 17, exit at Glen Avenue where you will see the spire of the Old Paramus Reformed Church ahead of you. Bear right on

This old schoolhouse in Ridgewood is maintained by the Paramus Historical and Preservation Society. (Author's own photograph)

East Glen Avenue and the schoolhouse will be on your left. Driving north on Route 17, exit at Linwood Avenue. Make a U-turn over Route 17 and, at the second traffic light, go right on Pleasant Avenue about three-fourths of a mile until you dead end on East Glen Avenue. You turn right again onto East Glen Avenue and the schoolhouse will be on your right.

Open: Wednesday 2:30 P.M.–4:30 P.M.

               Sunday 3 P.M.–5 P.M.

               Also by appointment

Admission: Free

     (R)

The tour is guided and lasts about 30 minutes. There is a three-page pamphlet describing the contents of the schoolhouse available at no charge. Booklets on the area, notepaper, and souvenirs are for sale. Nearby is the Duck Pond, which has picnicking facilities and grills, as well as recreational facilities.

While here take a look at the Old Paramus Reformed Church, where it is thought that George Washington once worshipped in 1778. The present edifice dates from about 1800 and was remodeled in 1873.

RINGWOOD

Ringwood Manor (c. 1810, with later additions)
Ringwood State Park

The location of this unusual house has been the site of the ironmaster's residence for well over two centuries, as iron ore was extracted and smelted nearby from 1740 until 1931. The present structure was built in 1810 by the ironmaster at that time, Martin Ryerson. He had purchased the iron works in 1807 and proceeded to tear down the existing residence, replacing it with what is now the section on your left as you face the front of the present manor house.

In 1853 Peter Cooper bought the ironworks and manor house. His daughter, Amelia, married Abram Hewitt in 1855, who was to become the last of the ironmasters. The manor house, as you see it today, reflects the changes and additions that were made by the Coopers and Hewitts. They appended outer buildings to the then-existing house, and later added a roof and stuccoed the house in order to give this essentially ungainly residence some symmetry. The well known architect Stanford White designed the porte cochere in the early 1900s.

Ringwood Manor and its ironworks have some historic

Located near the iron works, which were in operation from 1740 until 1931, Ringwood Manor was the ironmaster's home. (Courtesy N.J.D.O.E.P.)

significance. Robert Erskine, the ironmaster at the time of the Revolutionary War, became Geographer and Surveyor General to Washington. It is said that General Washington visited here on many occasions. As a major iron producer, Ringwood was important to the colonists in the Revolutionary War and to the Federal Government in later conflicts. The Cooper-Hewitt family generously left Ringwood Manor and its mostly Victorian furnishings to the state in 1936.

You enter a large hall with elaborately carved oak panelling in the Gothic Revival style, which was crafted in the late 1870s. The hall also has exhibits of guns dating to

This drawing room at Ringwood Manor features Louis XVI-style furniture Numerous other styles can be seen in the additional 20 rooms open to the public. (Courtesy N.J.D.-O.E.P.)

the Revolutionary War. The music room still has wallpaper with hand-painted seascapes, which was hung in 1867.

Downstairs, in the oldest section of the house, two remodeled parlors and a hall contain some handsome Dutch marquetry antiques. Unfortunately, these pieces have been neglected and desperately need repair.

Upstairs there are numerous bedrooms, the most interesting being Peter Cooper's, which contains mostly Empire period furniture. The mahogany dressing table was made by Abram Hewitt's father, John Hewitt, a well-known cabinetmaker.

Returning downstairs you will see the dining room with handsome cherry panelling. The Gothic Revival-inspired chairs and Empire sideboard were made by Peter Cooper.

There are also several rooms with exhibits of early Colonial kitchen equipment, Revolutionary and Civil War artifacts, and various pieces of metal work that were made at the Ringwood ironworks.

A visit here is recommended in order to see how a well-to-do family lived in the latter half of the nineteenth century in comfortable surroundings. Also, stroll around the gardens, which have products made by the old forge and columns and statues brought here from demolished buildings in New York and elsewhere. The State Park also offers numerous forms of recreation.

DRIVING DIRECTIONS: From Pompton Lakes take Route 511 north ten miles and then go right at the sign to Ringwood and Sloatsburg, New York. You continue four miles on this road. The entrance to the Manor will be directly ahead of you before the road veers to the right.

Open: Monday–Friday 10 A.M.–4:30 P.M.
             Saturday and Sunday 10 A.M.–5:30 P.M.
             Closed November through April
Admission: Adults—25¢
      (R)   Children—free
             Parking is 50¢

The tour is self-guided, aided by posted placards describing the contents of each room. You will see about 20 rooms (of the 78 in the house), and the tour lasts about one hour. Guided tours can be arranged in advance. A booklet entitled *The Story of Ringwood Manor* is available for 25¢. Postcards are also for sale. Art shows are held in the nearby barn from May 1st to October 1st.

Note: Ringwood State Park and adjacent Shepherds Lake offer a wide range of recreational facilities: swim-

This is an aerial photograph of Skylands, a meticulous reproduction of an English Tudor castle in northern New
Jersey. (Courtesy N.J.D.O.E.P.)

ming, boating (rowboats may be rented or you may
launch your own boat), fishing, nature center and
trails, trap and skeet shooting, picnicking, fireplaces,
and many more. There is also a snack bar that is
open daily during the summer and on weekends
thereafter. A restaurant, The Lodge, serves lun-
cheons and dinners all year.

Skylands Manor House (c. 1924)
Ringwood State Park

Strictly speaking this is not a historic house as it dates

only from the 1920s. However, it is a meticulous reproduction of an English manor house. It is partially half-timbered and essentially Tudor in style with a crenellated tower. This 44-room mansion was built by a very wealthy man, Clarence M. Lewis, as a summer house about 1924. Lewis spent a considerable amount of money to construct his manor, even to the extent of following the building techniques of English craftsmen of the sixteenth and early seventeenth centuries.

The interior of the manor house possesses magnificent hand-carved woodwork. Some of the panelling was brought here from old mansions in Europe and some was carved here for the house. As for the former, some entire rooms

This interior photograph shows the fabulous wood panelling for which Skylands is well known. (Courtesy David N. Poinsett)

This is the entrance to Skylands, which is now a New Jersey Historic Site. (Courtesy N.J.D.O.E.P.)

of panelling were imported for installation here. For example, the dining room boasts antique woodwork from England where it had been in a manor house dating from the first quarter of the seventeenth century. The delicate and detailed hand carving is not to be missed.

A college bought Skylands in 1953, and the state of New Jersey purchased the property 13 years later. At this writing there is no furniture in Skylands, but it is hoped that the state will furnish it in time.

This is another interior view of the richly decorated Skylands. (Courtesy David N. Poinsett)

Short of a trip to England, this is a grand way to visit an English manor house. While there, be sure to stroll through the beautifully landscaped gardens.

DRIVING DIRECTIONS: Same as for Ringwood Manor, but the entrance to Skylands is about two miles before the entrance to Ringwood, on the Ringwood-Sloatsburg Road. The sign will be on your right heading north.

Open: By appointment only, or ask at Ringwood Manor

Admission: Free

   (R)

The tour is guided and lasts about 25 minutes. A six-page booklet describing the manor house and its gardens is available at no charge.

RIVER EDGE

Von Steuben House (c. 1695, with additions 1752)
Main Street

This two-story Dutch colonial house was built about 1695 and consisted of the present parlor and hall downstairs and two bedrooms upstairs. In 1752 additions were made by extending the parlor and adding the kitchen, more bedrooms upstairs, and a gambrel roof. The resulting brick and fieldstone structure is what we see today.

This house has an interesting history. It was confiscated by the government during the Revolutionary War because the owners of the house, the Zabriskies, were Loyalists. It was then presented to Baron Von Steuben in 1783 in gratitude for his assistance to Washington in the War.

Downstairs one sees an attractive Colonial kitchen that has a good collection of kitchen equipment. Among the early American antiques in the large parlor is a handsome barrel-back chair dating from 1700.

There is a Colonial bedroom upstairs as well as two rooms that house permanent collections of Indian artifacts, dolls, toys, and relics, and which are also used for the display of rotating exhibits.

DRIVING DIRECTIONS: Take Route 4 west about four miles

This Dutch Colonial house was confiscated during the Revolutionary War because its owners were loyalists. Today Von Steuben House is owned by the state of New Jersey and is mostly furnished by the Bergen County Historical Society. (Courtesy N.J.D.O.E.P.)

from the George Washington Bridge and exit at River Edge—Business District. Bloomingdale's will be on your right. Follow the signs for the River Edge—Business District north on Route 503 about one-half mile. At New Bridge Road turn right, and the house is a short distance on your left. There are signs, but unfortunately, overzealous pop-art collectors or vandals keep removing them.

Open: Tuesday—Saturday 10 A.M.–12 P.M., 1 P.M. —5 P.M.

             Sunday 2 P.M.–5 P.M.

Closed Monday, Thanksgiving, Christmas,
and New Year's Day

Admission: Adults—25¢

(R)   Children under 12—free

The tour is guided and lasts about a half an hour.
Among the items for sale are maps, Revolutionary War
and Civil War souvenirs, and stationery.

SALEM

On Market Street, one of the main thoroughfares of this South Jersey town, there are some interesting old buildings in addition to the Alexander Grant House (c. 1721) and the Rumsey House (early 1700s). Across the street is the First Presbyterian Church, which was founded in 1821 and whose present Victorian edifice dates from 1854. Adjacent to the church is a house in the Federal style (1802), which is now used as a county office. At the junction of Broadway and Market is the old red brick Court House, initially dating from 1735. It was rebuilt in 1817 and again in 1908. This building is in disrepair, and its fate is undecided.

Continuing this informal tour of old Salem, go right on West Broadway a little way to the Friends Burying Ground. There you will see a most graceful sprawling oak tree that is about 500 years old. In the fall when the leaves on the tree are ablaze, the site is magnificent. It was under this tree that the founder of Salem, John Fenwick, made peace with the Indians in 1675.

To conclude the walking tour of the town, go left on East Broadway (from Market) one block until you see the brick Friends Meeting House, dating to 1772.

Outside of Salem proper there are some lovely eigh-

In Salem this is the Alexander Grant House with the Rumsey Wing, which is owned and maintained by the Salem
County Historical Society. (Courtesy David N. Poinsett)

teenth- and nineteenth-century houses that are private and
not open to the public.

Note: The Salem County Historical Society has a biennial

tour of houses in late April. The Society has prepared a pamphlet describing the noteworthy features of the eighteenth- and nineteenth-century buildings on that tour, and it may be used as a guide for your informal walking and driving visit.

**Alexander Grant House (c. 1721, with later addition c. 1728)
with Rumsey Wing (early 1700s)
79–83 Market Street**

These two "row type" houses are good examples of early brickwork in the Flemish bond style with pent roofs. The earlier house is to your right as you face them. Doorways have been made between the two houses so you move freely from one to the other.

The Salem Historical Society is most fortunate to have a notable collection of antiques, paintings, utensils, dolls, and china from Colonial through Victorian times. A few items of particular interest are: a tall case clock made by Thomas Wagstaff in London and brought to this country sometime before the Revolutionary War (when it was confiscated by the British who probably recognized the outstanding workmanship; later it was returned to the family) ; another English piece, a Jacobean oak chair dated 1630 (brought to this country in 1678) ; a prized Sheraton sideboard made and signed by George Whitlock, a Wilmington, Delaware, cabinetmaker; and a collection of the locally made Wistarburg glass (1739–1780) consisting of window panes, bull's-eyes, and glassware. The fireplaces and mantels throughout are either restored or period pieces that have been brought in from similar houses.

In the back is a new barn constructed in 1958 of old fieldstone, which contains farm tools and Indian artifacts. Also to the rear of the house is the small octagon-shaped law office built of brick about 1735.

Don't miss a visit to the Salem County Historical Society.

This tiny octagon-shaped law office is behind the Salem County Historical Society. (Courtesy David N. Poinsett)

DRIVING DIRECTIONS: Take Exit 1 (Delaware Memorial Bridge) on the New Jersey Turnpike and then drive southeast on Route 49 about nine miles to the center of

There is a well-appointed parlor in the Rumsey Wing of the Alexander Grant House. (Courtesy David N. Poinsett)

The dining room in the Alexander Grant House features this beauty of a sideboard. (Courtesy David N. Poinsett)

Salem. Market Street is the road to your left at the center
of town, and is easily located. Or, alternatively, from
Exit 1 follow Route 540 until it meets Route 45, which
is Market Street. The houses are on your right just before the road ends in the center of Salem.

Open: Tuesday through Friday, 1 P.M.–3 P.M., and
             by appointment
Admission: Adults—75¢
    (R)   Children—25¢
             (Adult groups of 25 or more—50¢ each)

The tour of the 12 rooms and barn is guided and lasts
about one hour. An eight-page brochure describing the
contents of the buildings is available free. Stationery and
postcards of the houses and historical and genealogical
pamphlets are for sale.

Hancock House (c. 1734)
Hancock's Bridge

This house is architecturally interesting as a good example of the Flemish bond style, being made of red and
blue glazed brick laid in an intricate pattern. The Quakers
of this area were particularly known for building houses
using such complex designs. As was customary in houses of
this style, the initials of the owner and his wife and the
year the house was completed are patterned into the brickwork. In this instance you will see on the gable the initials
"W & S H" for William and Sara Hancock and the date,
1734.

The interior of the house possesses its original mantels
and fireplaces. An unusual feature is the (then rare) closet
in the "keeping room" (where you enter) with the carved
grill above for ventilation. This detailed ornamentation
was hand-carved of separate pieces of wood that were then
fitted together. This ventilating panel is also characteristic
of South Jersey houses.

The furnishings are mostly of the Colonial period, with

Hancock House near Salem was the scene of a terrible massacre during the Revolutionary War and is now a New Jersey Historic site. It provides an excellent example of Flemish bond brickwork. (Courtesy N.J.D.O.E.P.)

two rooms downstairs, two bedrooms on the second floor, and exhibits of tools and farm and home implements on the third floor.

Another interesting feature of many houses in this area, which can be seen here, is the saddle-door, or hearse-door, which opens onto the rear of the house and is ahead of you (behind the sales desk) in the keeping room. Built above ground with no stairs, one could open the door and mount his horse. Or on the occasion of a death, the casket could be moved from the keeping room directly into the horse-drawn carriage.

Historically, Hancock's house is remembered as the scene of a massacre during the Revolutionary War. Early on the morning of March 21, 1778, 30 Quaker patriots, including Judge William Hancock, were massacred by British troops, which are believed to have numbered about 300 strong. Blood stains are still visible in the attic. This is the only house in New Jersey where such a massacre took place.

Also on the property is an old cedar cabin, recently moved here from a nearby location in Salem County. Dating from about 1640 to 1645, this unfurnished cabin should be looked at to appreciate the craftsmanship of the Swedish settlers in the "tongue and groove" dovetailing of the planks.

DRIVING DIRECTIONS: From the center of Salem take the main street, East Broadway, east to a fork in the road, where you bear right at the sign that says Hancock Bridge and follow this unnumbered road about four miles. The house will be on your right.

Open: Tuesday—Saturday 10 A.M.—12 P.M., 1 P.M. —5 P.M.

Sunday 2 P.M.–5 P.M.

Closed Monday, Thanksgiving, Christmas, and New Year's Day

Admission: Adults—25¢

(R)   Children under 12—free

The informed attendant gives a lecture of about ten minutes and then you are free to walk around the house. The entire visit lasts about 30 minutes. Postcards, notepaper, books on the county and state, and souvenirs are for sale.

Fort Mott State Park

Several miles north of Salem (or south of the Delaware Memorial Bridge exit on the New Jersey Turnpike) off Route 49 is this state park, which has picnicking facilities and fireplaces (charcoal can be bought there), playing fields, and playgrounds. The old Fort was built in the late 1830s to protect Philadelphia. The present one, which replaced it in 1896, affords a splendid view of the Delaware River Valley. Near the Fort is a burial ground for Confederate and Union dead. There is also a wildlife sanctuary.

SMITHVILLE

Although this is not a furnished historic house, as are most of the others in this book, it is included because it is an interesting place to visit. A rambling stage coach inn (whose original section dates from about 1787), now a restaurant, is the nucleus for the development of this highly successful Historic Smithville complex. The enterprising "possessors," as the owners, Mr. and Mrs. Fred Noyes, call themselves, have brought in numerous, old South Jersey buildings dating from as early as the late eighteenth century. As a result, one may observe the architectural style of the various buildings, such as the extant inn, sweet shop, clam house, and grist mill. Their respective products and others may be purchased at each.

The most exciting development at Smithville is the plan for a Historic Village Reconstruction Museum, which will represent life in a typical South Jersey community in the 1850s. Old, tired buildings in disrepair (mostly dating from 1820 to 1870) are now sitting on the proposed site awaiting reconstruction and refurbishing. In striking contrast there are a few buildings already completed, such as Hewitt House, which dates from about 1850. Final details of this museum village have not been decided, but most likely there will be an admission charge and restricted

191

hours. The houses and buildings will be furnished, and crafts will be demonstrated. Plans call for opening part of the village in the near future.

DRIVING DIRECTIONS: Heading south on the Garden State Parkway, take Exit 48. Then drive south on Route 9 for four miles, where you will find yourself in the middle of the village. Heading north on the Garden State Parkway, you have to exit at 50, then turn around and get back on the Parkway, and head south until Exit 48.

Open: Daily. You can actually walk around the village anytime, but the shops open around 11 A.M.

(R)

At the Historic Village that is now under construction in Smithville, this building, Hewitt House, has had its face lifted.

This shows an old house at Smithville waiting its turn to be refurbished.

Note: Besides the original Smithville Inn, this restaurant and village complex has other places to eat and a multitude of gift items for sale, including postcards, notepaper, and books on Smithville.

SOMERS POINT

Somers Mansion (c. 1726)
Shore Road at the Circle

This is one of the oldest extant houses in Atlantic County, dating from 1726 or before. The two- and a half-story dwelling with a second-story balcony is notable for its unusual arched roof, which resembles the hull of a ship upside down. The builder of this house was a ship builder, and he must have possessed foresight, as his grandson, Richard Somers, was to become a Naval hero during the War with Tripoli in 1804.

One of the four rooms visited has its original fireplace and handsome panelling. Four chairs and a chest-on-chest in the Chippendale style that belonged to the Somers family are also seen.

Note: It has been necessary to close Somers Mansion to do some repair work, primarily on the distinctive roof. The state hopes to reopen it soon.

DRIVING DIRECTIONS: Take Exit 30 off the Garden State Parkway. Proceed east on Route 52 for a mile and a half where you will meet a traffic circle. Drive almost all of the way around the circle and the house will be on your right.

194

Somers Mansion near the Jersey shore is known for its distinctive roof. Looking at the roof it appears to be a ship upside down. (Courtesy N.J.D.O.E.P.)

Open: Tuesday–Saturday 10 A.M.–12 P.M., 1 P.M.–
               5 P.M.
               Sunday 2 P.M.–5 P.M.
               Closed Monday, Thanksgiving, Christmas,
               and New Year's Day
               (Planned reopening July 1971)
Admission: Adults—25¢
       (R)   Children under 12—free
The tour is self-guided, aided by posted descriptions of the contents, and lasts about 15 minutes. There are postcards, notepaper, and books of the house and New Jersey for sale.

SOMERVILLE

The Old Dutch Parsonage (c. 1751)
65 Washington Place

Built with bricks brought (probably as ballast) from Holland in 1751, this house is unique in that the smokehouse is on the third floor, having been moved there from its outside location to discourage foraging by troops during the Revolutionary War. Looking at the front of the house you can see where a porch was added and later removed, as evidenced by the discoloration of the bricks between the first and second stories.

The builder of the house was the Reverend John Frelinghuysen, a Dutch Reformed minister, who started a seminary here that later became Rutgers Theological Seminary. When he died, his widow married the Reverend Jacob Hardenberg, who, in 1785, became the first President of Rutgers University in New Brunswick.

The interior needs refurbishing and has been temporarily closed. It will be reopened soon, hopefully in 1971. The five rooms visited have Colonial furnishings. In the living room is a handsome gilded Dutch mirror that Mrs. Freylinghuysen brought from Holland in the early 1750s.

DRIVING DIRECTIONS: From the traffic circle in Somerville

In Somerville one can visit the Old Dutch Parsonage, which is a New Jersey Historic Site. (Courtesy N.J.D.O.E.P.)

(where Routes 206, 202, and 28 converge) take Route 206 south a few blocks to the first traffic light at Somerset Street, where you turn left. Then turn right at the next corner, Middagh Street, and then take your next left onto Washington Place. The parsonage is on your right, and Wallace House is a few doors down on the left.

Open: Tuesday–Saturday 10 A.M.–12 P.M., 1 P.M.–
            5 P.M.
            Sunday 2 P.M.–5 P.M.
            Closed Monday, Thanksgiving, Christmas,
            and New Year's Day

(See remarks about temporary closure above.)

Admission: Adults—25¢

(R)   Children under 12—free

The tour is guided and lasts about 15 minutes. A one-page write-up on the house is available at no charge. If the caretaker is not there, go across the street to Wallace House or to her residence, which is next door to Wallace House. Notepaper, postcards of the house, books on New Jersey, and souvenirs are for sale.

Wallace House (c. 1778)
38 Washington Place

General George Washington and his wife spent the winter of 1778–1779 here, at which time this wooden house was still under construction. It was here that he planned the Indian Campaign of 1779.

Among the items relating to the Washingtons' residency in the house are the Queen Anne-style dining room table, said to be a gift from their neighbor and friend, Reverend Frelinghuysen; some pieces of a Lowestoft tea set that belonged to Martha Washington; a pair of iron forceps used to extract one of the General's teeth that winter; and a large campaign chest.

The house has period furnishings, and the State of New Jersey is doing some restoration of the interior. The parlor and two upstairs bedrooms were recently painted, and the picture molding (a later addition) was removed to reveal the original Grecian, or Greek Key, molding.

Fortunately the house has its original floors (except in the dining room). There is an attractive kitchen with late eighteenth-century utensils. The bedrooms upstairs are furnished in the period. A unique feature of the house is the built-in closets, an innovation at that time.

Since the tour enters and leaves the house through the

Located across the street from the Old Dutch Parsonage is Wallace House. During the winter of 1778–1779 George Washington and his wife stayed here. (Author's own photograph)

This is the attractive parlor at Wallace House, which has recently been refurbished. (Courtesy David N. Poinsett)

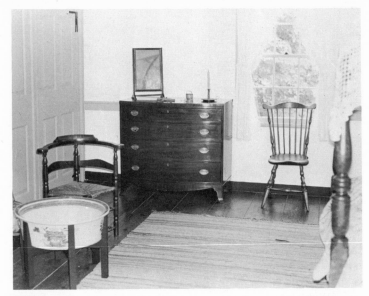

A part of one of the bedrooms at Wallace House is seen in this picture. (Courtesy David N. Poinsett)

rear door, it is suggested that the visitor walk around to the front to see the main entrance.

See the Old Dutch Parsonage above for driving directions and opening hours, which are the same. The guided tour lasts 30 minutes.

Note: The Duke Gardens Foundation is located nearby on Route 206, a few miles south of the intersection of Routes 206, 202, and 28. There are 11 lovely gardens, each indigenous to a different country or period. A brochure describing the gardens and giving opening times is available from the Foundation.

SPRINGFIELD

Cannon Ball House (c. 1760)
126 Morris Avenue

This one-time farmhouse was one of the four houses that survived the Battle of Springfield on June 23, 1780. It did receive some damage, however, and on the west side of the house one can see where a cannon ball struck this building during the Battle.

Inside there are some original moldings around the ceilings together with a handsome built-in cupboard that dates from the construction of the dwelling in about 1760.

This is the headquarters of the Springfield Historical Society, and there are exhibits of Revolutionary and Civil War artifacts and items pertaining to local history.

DRIVING DIRECTIONS: Northbound, take Exit 140 on the Garden State Parkway. Then drive west on Route 82, which becomes Route 24 (which in turn is Morris Avenue). The house is about two miles ahead on your right. Southbound, take Exit 143 and follow Route 24 west about four miles until you see the house on your right.

Cannon Ball House is one of the four houses in Springfield that was not destroyed during the Battle of Springfield in 1780. It is the headquarters of the Springfield Historical Society. (Author's own photograph)

Open: Sundays from September 1st through June
                  30th, from 2 P.M. to 4 P.M., or by appoint-
                  ment
Admission: Free
           (R)

The tour is guided and lasts about 15 minutes. Available free of charge is a brochure, *Famous Landmarks of Historic Springfield,* which the visitor may use as a guide for a walk or drive to view the 20 noteworthy churches, houses, and bridges. Plates, tiles, and postcards of Cannon Ball House and a booklet describing the Battle of Springfield are for sale.

STANHOPE

Waterloo Village Restoration (c. 1740 with later buildings added through the 1850s)
R. D. (Stanhope)

This village was settled by the English about 1740 as a farming community. However, it wasn't until 1763 when an iron forge was constructed that Waterloo began to prosper. During the Revolutionary War the iron produced here was used for gun barrels for the Colonial troops. After the end of the war, the iron forge stopped production for the lack of wood, which was required for fuel.

Prosperity returned to the village with the building of the Morris Canal beginning in 1824. Waterloo was one of the main depots along the canal, which flourished in the 1860s. With the rise of rail traffic after the Civil War, the use of the Morris Canal began to wane. It was finally closed in 1924. Having lost its function as a transportation center, Waterloo became nearly a ghost town until Messrs. Percival Leach and Louis Gualandi became enchanted with the old village several years ago and gradually bought up most of the existing buildings. They have admirably recreated an Early American village and its subsequent

The Waterloo Village Restoration is north of Stanhope. This village went through many stages, but experienced its most successful times when it was a main stop on the Morris Canal. This is an exterior picture of the Canal House, which was used as workmen's living quarters. (Courtesy David N. Poinsett)

In the Canal House one visits this attractive dining room. (Courtesy David N. Poinsett)

nineteenth-century development as a transportation center on the Morris Canal.

There are 11 buildings open to the public. However, only four fall within the scope of this book:

Canal House (c. 1760). This three-story stone dwelling was built as living quarters for the forge's workmen. There is a dining room downstairs and another one on the second floor together with a parlor. These rooms are tastefully furnished with antiques and china from Colonial through Victorian times.

Wellington House (c. 1859). This is a good example of a Victorian-style house with "gingerbread" decoration, a turret, and several porches. Only the downstairs is open to the public, and all three rooms are used as dining rooms. The marble fireplaces are original. These rooms have silver pieces, Venetian glass, and a handsome Chippendale linen

This is the Wellington House, a good example of Victorian architecture in Waterloo Village. (Courtesy David N. Poinsett)

This is the elegant dining room in Wellington House. (Courtesy David N. Poinsett)

press on display. On the front porch observe the Victorian bird cage.

The Homestead (c. 1750s with addition c. 1860). Prior

Next to the Wellington House is the Homestead. This structure was first a barn and then converted into a house. (Courtesy David N. Poinsett)

to the Revolutionary War this was a barn, and at a later date it was converted into a house. The four rooms visited contain both antique and non-vintage furniture.

Old Stage Coach Inn (c. 1740 with addition c. 1830). Facing this inn, the section to your right is the oldest, though this is not discernible because of the Greek Revival facade, which was added later. During the heyday of the Morris Canal this inn and tavern were well known. In the parlor there is a handsome English piano dating from 1806 that is still playable. There are several dining rooms tastefully furnished with Colonial antiques. Upstairs there are numerous bedrooms, the only ones viewed in the Village. Note the colorful stencilling, which was so popular in Colonial times, at the top of the walls below the ceiling. In one of the bedrooms there is an interesting Early American four-poster bed with finials carved in the form of pineapples, a symbol of hospitality. If the finials were removed during his visit, a guest would know that he had overstayed!

The other buildings are primarily devoted to exhibits and demonstrations of Early American crafts. On Saturdays and Sundays the crafts of the gunsmith, cabinetmaker, candlemaker, blacksmith, and weaver are demonstrated. Also visit the Fragrant Herb and Apothecary Shop and herb garden, as well as the grist mill and church (c. 1859).

The Old Stage Coach Inn at Waterloo Village was a popular stopping place during the heyday of the Morris Canal. (Courtesy David N. Poinsett)

A walk along the picturesque Morris Canal is most enjoyable. Both adults and children will enjoy a visit to the Village.

DRIVING DIRECTIONS: Routes 46, 10, 206, and Interstate 80 all meet in Netcong. Proceed north on Route 206 from this intersection about two miles, where you will see a marker to the Village. You then go left on a country road about three miles to the Village.

Open: Weekends 11 A.M.–6 P.M. (or by appointment) April–June, September–November

               Tuesday–Sunday 11 A.M.–6 P.M. July and August
               Note: The last tour tickets are sold at 3:30 P.M. weekdays and 4:30 P.M. weekends

Admission: Adults—$2.50
               Children—75¢
               Students—50¢ (in groups)
               Saturday and Sunday—Purchase tickets at the main entrance gate and then walk around the Village at your own pace. Costumed guides or craftsmen will be on hand to give explanations in each of the buildings.
               Tuesday–Friday—Go directly to the General Store where you will find a guide that will conduct a tour. It is advisable to call ahead during the week and make an appointment: (201) 347-0900.

The tour lasts about two hours. The General Store sells many things: penny candy, old relics, postcards, booklets on the Village and its crafts, and maps of the Morris Canal. There is a six-page brochure describing the Village, which is available at no charge in the General Store. From Memorial Day through Labor Day there is a grill where you can buy hamburgers, hot dogs, and other snacks. Picnicking is allowed anywhere in the Village. There are no group tours given on Saturday and Sunday. The Village offers a unique feature for groups. Besides a guided tour, luncheon will be served in the Wellington House in lovely surroundings. Contact the Village for menus, prices, and arrangements.

Note: In the summer, concerts are given Saturday nights under a tent. (I noted that Van Cliburn and Pablo Casals have performed here.) The price of the concert ticket entitles the holder to visit all the buildings that are kept open on such occasions.

TRENTON

The Old Barracks (c. 1758–1759)
South Willow Street

This is the only remaining one of five barracks that were erected by orders of the Colonial Assembly for the purpose of quartering troops during the French and Indian War (1755–1763). During the Revolutionary War Hessian, British, and American soldiers were all quartered here at various times. It was also used as a hospital. At the conclusion of the war, the barracks were converted into row houses.

As you face this U-shaped structure, made of undressed fieldstone with lovely porches around it, look at the center section toward the right, and you will note that the stone in this section is different from the rest. Interestingly, in 1813 this 40-foot section was torn down to make way for a new road to the State Capitol. It was subsequently replaced when the old barracks were restored during the 1910s.

The visitor should be advised that the barracks are not all furnished as they were during the latter part of the eighteenth century. However, the Old Barracks Associa-

In Trenton one may visit the Old Barracks, the remaining barracks of five originally built during Colonial times. (Courtesy David N. Poinsett)

Many of the rooms at the Old Barracks are handsomely furnished with antiques like this one. (Courtesy David N. Poinsett)

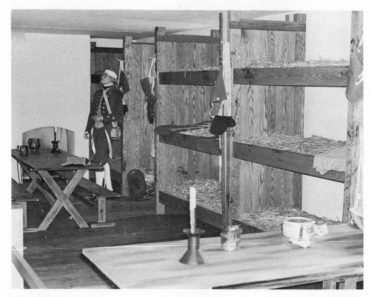

This reconstructed "squad room" shows the living quarters of soldiers in the Old Barracks. (Courtesy David N. Poinsett)

tion has now reconstructed a "squad room," which shows the barren living quarters that the soldiers experienced with rows of bunk beds (newly built to old specifications) and a simple table with benches at which they ate their meals.

The interior consists of mostly smaller rooms, which resulted from the conversion of the barracks to row houses, and most of these rooms are handsomely furnished by the various organizations that maintain the rooms. There is a great wealth of Colonial, Revolutionary, and Federal period antiques, and some English pieces. A most unusual piece is a mahogany writing table and desk with cabriole legs that was made for Aaron Burr in 1800.

In the rooms you will also find displays of Chinese export porcelain, silver, quilts, and Colonial kitchen equipment. There are also exhibits of Continental currency, swords, and muskets, together with a narration and dioramas of the Old Barracks and of the Battles of Trenton and Princeton (December 1776–January 1777).

DRIVING DIRECTIONS: Take Exit 7 on the New Jersey Turnpike, and then follow Route 206 (which becomes Broad Street) seven miles to the center of Trenton. At the intersection of Broad and Market Streets, turn left onto Market. After the first traffic light at Warren Street, Trent House (see below) will be immediately on your left. If you continue on this road, you will be in the midst of the capitol complex. The Old Barracks and the Old Masonic Lodge (see below) will be on your right beneath the gold dome of the Capitol. Because of construction, it is difficult to be more specific.

Open: Monday–Saturday 10 A.M.–5 P.M., June–August

Monday–Saturday 10 A.M.–4 P.M., September–May

Sundays 2 P.M.–5 P.M. all year

Closed Thanksgiving, Christmas, New Year's Day, and February 22nd

Admission: Adults—75¢

(R)   Students—50¢

Children through 12—25¢

Maximum charge for family—$2.00

The tour is guided by costumed, informative hostesses, and lasts about 45 minutes. A four-page brochure on the Old Barracks is available free. An expanded booklet on the Old Barracks, ash trays, plates, postcards, and other souvenirs are for sale.

Old Masonic Lodge (c. 1793)
South Willow Street

For Masons, this Georgian-style "lodge," built of field-

In Trenton one may visit the Old Masonic Lodge. (Courtesy David N. Poinsett)

stone, is thought to be one of the oldest Masonic Temples in the United States. This restored lodge has exhibits on Masonry downstairs. Upstairs is the lodge room with some of the original Windsor-style chairs that were made for the lodge in 1793.

DRIVING DIRECTIONS: Same as for the Old Barracks. The Lodge is located across the street from the Old Barracks and down a few steps.

Open: Monday–Friday 10 A.M.–12 P.M., 2 P.M.–
             4 P.M.

Admission: Free

The tour is guided or self-guided and lasts about ten minutes. A four-page booklet is available free.

William Trent House (c. 1719)
South Warren Street

This two-story red brick structure with white cupola was completed in 1719 and is an excellent example of an Early Georgian mansion. It was built by the city's namesake, William Trent. While portions of the house have been restored, about 80 percent is original. Important features of the house's interior are the staircase, fireplaces, and inside shutters.

An inventory of the house's contents in 1726, drawn up in connection with the settling of William Trent's estate, has made it possible to furnish the house as it was then.

Trent House was built by the namesake of Trenton, William Trent. The house is an outstanding Early Georgian structure. (Courtesy N.J.D.O.E.P.)

The dining room of Trent House features some handsome William and Mary chairs. (Courtesy David N. Poinsett)

The quality furniture is of the William and Mary, and Queen Anne periods, of British and American origin. The grandfather's clock in the downstairs hallway be-

This bedroom at Trent House is furnished with quality Queen Anne and William and Mary antiques. (Courtesy David N. Poinsett)

longed to William Trent's son. This piece, dating from 1760, is the only furnishing that was owned by the Trent family.

Take note of the massive Delaware Valley cupboard or "kas" in the upstairs hallway. This oak piece is decorated with panelling. See if you can determine the hiding place in the cupboard for valuables.

A visit to this handsome and richly furnished house is recommended.

DRIVING DIRECTIONS: For directions, see the Old Barracks above. The Trent House is in the heart of the area of downtown Trenton that is being rebuilt. It is located just south of the new State buildings, and there are posted signs in central Trenton pointing the way.

Open: Monday–Saturday 10 A.M.–4 P.M.

                Sunday 1 P.M.–4 P.M.

                Closed Thanksgiving, Christmas, and New Year's Day

Admission: Adults—25¢

                Children—10¢

A free, four-page brochure on the house and William Trent is available. A larger book on the same topics may be purchased for $1.00. Other items may also be purchased, and perhaps the most interesting are pewter spoons, some of which are cast from a mold found in the house. The tour is guided and lasts about 40 minutes.

The Isaac Watson House (c. 1708)
151 Westcott Street

Isaac Watson built this two- and a half-story farmhouse in 1708. This date may be seen on the inscribed stone on the second story above the entrance. The stone house has a steep roof with a pent roof between the first and second stories. It also has unusual leaded windows.

The interior possesses two of the original large fireplaces and some beveled beams. However, it has been necessary to restore the wall panelling. Downstairs the parlor and dining room/kitchen are furnished with pieces dating from 1790 or before, a criterion established by the Daughters of the American Revolution who administer the property. (The DAR has a 99-year lease from Mercer County.) There is a noteworthy painted cupboard of Pennsylvania Colonial vintage, whose doors have all but one of the original window panes, and some William and Mary chairs, dating from about 1700.

Upstairs there are three furnished bedrooms. Among the interesting pieces is a linen press. Crewel fanciers should be sure to see the exquisite curtains, bedspread,

The oldest house in Trenton dates from 1708. Known as
the Isaac Watson House, it is maintained by the Daughters
of the American Revolution, who have a 99-year lease
from Mercer County. (Author's own photograph)

and dust ruffle that have been stitched in recent years to
Early American patterns.

DRIVING DIRECTIONS: From the center of Trenton proceed
south on South Broad Street (Route 206) for about two
and three-quarter miles to Park Avenue, where there is
a traffic light. Go right on Park Avenue and proceed for
six small blocks until you turn left on Westcott Street.
The house is a short distance on your right.

Open: By appointment only. Write to the above
             address or call (609) 888-2062.

Admission: Free

    (R)

The tour is guided and lasts about 20 minutes. A four-page brochure is available at no charge. A much-expanded version is for sale ($1.00) , together with postcards, plates, and napkins of the house.

Behind the house is Roebling Park, a wildlife sanctuary, where picnicking and charcoal grills are available.

WASHINGTON CROSSING

McKonkey Ferry House (c. 1930)
Washington Crossing State Park

On that stormy and cold Christmas night in 1776 when Washington and his troops crossed the Delaware River, they are said to have rested a while in a ferry house and tavern similar to this before marching on to Trenton, where they surprised and defeated the merry-making Hessian troops. This structure is a reproduction, built about 1930.

This white wooden structure has scalloped shingles on the front, and the interior has been reproduced as a typical Colonial tavern with the taproom and dining room downstairs and a bedroom upstairs. The furnishings are typical for an inn of that period, with Colonial kitchen utensils displayed in the dining room fireplace. The handsome dining room table is made of one slice of a trunk of a cherry tree.

DRIVING DIRECTIONS: In Washington Crossing State Park, which is about eight miles northwest of Trenton on Route 29.

Open: Tuesday—Saturday 10 A.M.—12 P.M., 1 P.M.
               —5 P.M.
               Sundays 2 P.M.–5 P.M.
               Closed Monday, Thanksgiving, and December 15th–January 1st, but it is open on Christmas afternoon for the colorful, annual re-enactment of the Crossing of the Delaware.

Admission: Adults—25¢
   (R)   Children under 12—free

A tour is guided and lasts about ten minutes. A four-page brochure is available at no charge. A booklet on the

After crossing the Delaware on Christmas Night in 1776, Washington and his troops may have stopped awhile at the McKonkey Ferry House before going on to Trenton. This is a New Jersey Historic Site. (Courtesy N.J.D.O.E.P.)

Battle of Trenton and other publications having to do with New Jersey and its place in history are for sale. Souvenirs, postcards, and notepaper may also be purchased.

Note: An entire day's outing can be planned around Washington Crossing State Park, where there are many things to see and do.

The Flag Museum, which is nearby, has a collection of early flags and several dioramas portraying Washington's crossing the Delaware and another of the Battle of Princeton. This is only open from June to October, from Tuesday–Saturday, 11 A.M.–5 P.M. and Sundays, 1 P.M.–5 P.M. Admission is free. (The Flag Museum hopes to be open longer when more staff is acquired.)

The Nature Center has guided or self-guided tours covering more than two miles of trails. The Nature Center also has a museum with exhibits of animals, birds, and flowers. It is open during the summer from 10 A.M.–4 P.M.

In the *Open Air Theater,* there are theatrical productions on Friday and Saturday evenings during the summer.

Picnicking facilities with grills and playgrounds are nearby. On the *Pennsylvania side of the Delaware* there is another extensive park where one can see a Durham boat (used for crossing the river by Washington and his men) ; a reproduction of the famous picture of "Washington Crossing the Delaware" by Luetze, together with a fifteen-minute recorded presentation about the event; and the Old Ferry House where, according to legend, Washington waited before crossing the Delaware (this has been restored and is now being furnished, the downstairs being nearly completed) . Further north on Route 32 (Pennsylvania) is the Thompson-Neely farmhouse (1701) , where Washington stayed and planned the strategy for the crossing and subsequent attack on Trenton.

WAYNE

Dey Mansion (c. 1740)
Preakness Valley Road
199 Totowa Road

This is an excellent example of a Georgian-style mansion with a gambrel roof built by the Dey family between 1720 and 1740. Facing the front, one can readily see how the appearance of this brick dwelling is enhanced by the use of brownstone around the windows, front door, and corners. Interestingly, on the other sides of the house, the basic walls are made of brownstone and fieldstone, with brick decoration around the doors and windows. This house is noteworthy historically since General George Washington spent the months of July, October, and November of 1780 here. In the aide's room one sees five of the ten pewter plates that were given by the Dey family to Washington on the occasion of his stay.

When this mansion was restored in 1934, every step was taken to restore it authentically, and the results are most gratifying. Some of the original pine floors, panelling, chair rails, and fireplaces exist. One of the outside window

226

In Wayne one can visit the Dey Mansion, an outstanding example of Georgian style architecture. It is owned and maintained by the Passaic County Park Commission. (Courtesy David N. Poinsett)

This picture of the back of Dey Mansion shows the interesting stone and brick work. (Courtesy David N. Poinsett)

The wide center hall in Dey Mansion contains noteworthy antiques and paintings. (Courtesy David N. Poinsett)

shutters was found and new ones were duplicated from it for the other windows.

On the first two floors there are eight rooms and two spacious center halls. The house is furnished with an excellent collection of eighteenth-century antiques, mostly representing the Chippendale and Queen Anne styles.

The third floor has an interesting collection of colonial artifacts, which children will particularly enjoy. Also, be sure to point out to them the trap door in the roof where Washington's soldiers peered out to see if the British were coming.

Stroll in the lovely gardens outside and wander around the barnyard with its live animals. Among the reconstructed buildings in the barnyard are a spring house and tool shed. A Colonial barn houses the animals. Further expansion is planned for the barnyard.

A visit to the Dey Mansion is highly recommended. By the way, the most informative curator is a tenth-generation Dey.

DRIVING DIRECTIONS: Heading north on the Garden State Parkway, take Exit 153B; heading south take Exit 154. After exiting, go west on Route 46 about four and a half miles. Drive north (right) on Riverview Drive about one mile, where Totowa Road veers off to your right. The Dey Mansion is a short distance on your left.

Open: Tuesday, Wednesday, Friday 1 P.M.–5 P.M.
          Saturday and Sunday 10 A.M.–12 A.M.,
          1 P.M.–5 P.M. or by appointment

Admission: Adults and children over 16—50¢
     (R)   Children under 16—free

The tour of the mansion is guided or self-guided and

This is one of the drawing rooms in Dey Mansion. One of the features of its antique furnishings is this outstanding highboy. (Courtesy David N. Poinsett)

The township of Wayne owns and maintains this Dutch Colonial farmhouse known as the Van Riper-Hopper House. (Courtesy David N. Poinsett)

lasts about 50 minutes. On Sundays there are teenage guides dressed in period costumes who conduct the tours.

There are two four-page pamphlets available free; one describes the restoration of the Mansion and the other gives a history of the Mansion with particular emphasis on Washington's stay.

There are several picnic tables available behind the Mansion. The Passaic County Golf Course is also adjacent. It and the Dey Mansion are owned by the Passaic County Park Commission.

Van Riper-Hopper House (c. 1786)
533 Berdan Avenue

This one-and-a-half-story, Dutch Colonial-style farm-

This is the attractive kitchen in the Van Riper-Hopper House. (Courtesy David N. Poinsett)

house, built of fieldstone, is in the process of being restored. The kitchen was recently tastefully completed in the style of the Revolutionary period. A cross section of the wall was left exposed to illustrate how structures were built in early times of a mixture of stones, straw, sticks, and mud called "wattle and daub." The parlor has recently been renovated in eclectic eighteenth- and nineteenth-century furnishings. Plans call for the opening of the upstairs sometime soon, probably in 1971. There is an herb garden outside.

DRIVING DIRECTIONS: From Route 46 (near Singac) take Route 23 about four miles until Route 202 branches off to the right. You then head north on Route 202 just until you take a right on Jackson Avenue, which runs into Route 504 (the Paterson and Hamburg Pike). You continue east for about three-quarters of a mile and then take a left on Berdan Avenue (also Route 502). The farmhouse will be on your left at the end of the reservoir.

These directions are not the most direct if you are traveling from the eastern sector of the state. Due to poor markings of the roads coming from the Garden State Parkway, the longer way I have outlined probably will prove easier to follow.

Open: Tuesday, Friday, Saturday, and Sunday
               1 P.M.–5 P.M.
               Closed Thanksgiving, Christmas, and New
               Year's Day

Admission: Free
           (R)

The tour is guided and lasts about 15 minutes. For 10¢ one can obtain a four-page pamphlet on the farmhouse. Also for sale are postcards and stationery of the house.

WEST ORANGE

Glenmont (c. 1880)
Llewellyn Park

This 23-room, gabled, Victorian mansion was built in 1880 of pressed red brick and half-timbered wood. The interior woodwork is particularly handsome as witnessed in the oak-quartered panelling in the entrance hall, the mahogany staircase, and the painted, recessed ceilings in the den and library.

Thomas Edison bought this rambling mansion on the occasion of his second marriage (in 1886) to the former Mina Miller. He resided here with his family until his death in 1931. He died in one of the many upstairs bed-rooms that are seen. It is indeed fortunate that the mansion still has the Edison family furnishings. The 12 rooms and two spacious halls that are shown on the tour contain Victorian furniture, paintings, silver, books, Venetian glass, and Edison memorabilia. Edison called the upstairs living room and library his "thought-bench," whence came ideas for his "work-bench" (or laboratory).

A tour of Glenmont is recommended as it affords you the two-fold opportunity to see how Thomas Edison lived

Thomas A. Edison, as photographed in 1917, is seen reading on the lawn in front of his Victorian home, Glenmont. Edison's home is owned and maintained by the National Park Service. (Courtesy U.S. Department of the Interior, National Park Service, Edison Historic Site)

and to view an excellent example of a well-preserved, Victorian-style mansion.

DRIVING DIRECTIONS: Take Exit 147 southbound or Exit 145 northbound on the Garden State Parkway and follow green signs to the Edison National Historic Site. The signs (profiles of Thomas A. Edison) will lead you two miles west on Park Avenue until it ends at Main Street, where you turn right. The Edison Headquarters is a short distance on your right.

Open: Monday–Saturday 10 A.M.–4 P.M.
             Closed Sunday and holidays

Admission: Adults—50¢ (includes a tour of Glenmont
and of the Laboratory)

Children under 16—free when accompanied
by an adult

Note: In order to tour Glenmont one must go to the Edison Headquarters at the Laboratory on Main Street
to obtain your ticket. Here the park attendant will
give you directions on how to drive to Glenmont,
which is located in the very exclusive residential

This was Edison's second-floor living room and contains
the Edisons' furnishings. You can see Edison's desk, which
he called his "thought bench." The ideas generated here
were experimented on in his "work bench" or laboratory,
which is located at the foot of the hill below Glenmont.
(Courtesy U.S. Department of the Interior, National Park
Service, Edison Historic Site)

area, Llewellyn Park. The guard at the entrance to the park will probably ask to see your ticket before allowing you to enter. No buses are allowed in Llewellyn Park, so if you are on a group tour, be prepared for a brisk walk uphill from the Edison Laboratory.

The guided tours begin promptly on the hour and last about an hour. If you are late in arriving at Glenmont, you are not allowed to join a tour already in progress. If you are planning to visit on a Saturday, it is recommended that you write ahead for a reservation and ticket, as each hourly tour is limited to 20 people, and Saturdays are very popular. It would be disappointing to drive a long way only to find out that all the tours are filled. Souvenirs, postcards, and books on Edison are sold at the Headquarters. An eight-page brochure describing Glenmont and the Edison family is provided at no charge.

For your information the guided tours of the Laboratory go all day from 9:30–4:30. This is a most interesting tour, and you will no doubt wish to visit the Laboratory as well as Glenmont.

WOODBURY

The Hunter Lawrence House (c. 1765, with later addition c. 1870)
58 North Broad Street

The rear section of this house was built in 1765, and the Victorian front portion was added in the 1870s. At one time it was the home of John Lawrence, brother of the Revolutionary War naval hero, Captain James Lawrence (see Burlington: Lawrence House).

Hunter Lawrence House is now the headquarters of the Gloucester County Historical Society. Although the interior design is primarily Victorian, in the older section there is a 200-year-old fireplace, panelling, and corner cupboards, which were rescued from other demolished houses in Gloucester County. While there is some furniture—notably a Jacobean oak chest brought from England in 1683, Elizabeth Haddon's Queen Anne chair and desk (see Haddonfield: Greenfield Hall), and some Victorian pieces—most of the rooms are devoted to numerous displays relating to the life and the history of the county, including Indian artifacts.

There are four additional rooms in the basement. One

In Woodbury one may visit the Hunter Lawrence House, which is owned and maintained by the Gloucester County Historical Society. (Courtesy David N. Poinsett)

features the fireplace before which Betsy Ross is said to have been married on November 4, 1773. The fireplace was then in a tavern in Gloucester County, which building

Inside the Hunter Lawrence House, one can see the chair and secretary that belonged to Elizabeth Haddon, namesake of Haddonfield. (Courtesy Gloucester County Historical Society)

has now been demolished. There is also a Colonial kitchen with English pewter plates and cutlery that were retrieved in the 1950s from a British ship that had been sunk in the Delaware River during the Revolutionary war.

Also on view at Hunter Lawrence House is the fireplace before which Betsy Griscom may have married John Ross. (Courtesy Gloucester County Historical Society)

DRIVING DIRECTIONS: Take Exit 3 on New Jersey Turnpike, and then take Interstate Highway 295 south to the Woodbury exit. You are now on New Jersey Route 45, which is also Broad Street. Proceed along Broad Street about two miles, and the house will be on your left, just before the center of town and the courthouse.

Open: Wednesday 2 P.M.–4 P.M. and by appointment. Closed July and August

Admission: Free
    (R)

There is no guide but displays are labelled. A two-page brochure on the history of the county is available. Notepaper and books on the history of the county may be purchased.

Whitall House (c. 1748)
Red Bank Battlefield, 100 Hessian Avenue

The house is located in a park on the site of the Battle of Red Bank, October 22, 1777. (Note: this is not near the present town of Red Bank.) It was the home of a heroine of that conflict, Mrs. Ann Whitall. Although her story may be apocryphal, it is endeared in local history. According to the legend, Mrs. Whitall was spinning in her house while the battle raged outside. When a cannonball hit the house, she calmly carried her wheel to the cellar and resumed spinning. After the battle, she cared for the wounded of both sides.

This is Whitall House at the Red Bank Battlefield. Ann Whitall supposedly continued her spinning in this house during the battle. (Courtesy David N. Poinsett)

Her spinning wheel and Bible are in the house. The house also contains some Colonial furniture, original fireplaces, and a built-in corner cupboard.

The date of the main portion of the house may be seen on the chimney. The small stone wing of the house was built by Swedish settlers earlier in the eighteenth century. It is not open to the public.

Note: Perhaps the most enjoyable way to visit the house would be to take a picnic lunch, since the park overlooks the Delaware River (which is to be a beneficiary of the state's new clean water program) . The Red Bank Battlefield Monument is within walking distance of the house.

DRIVING DIRECTIONS: Same as for Hunter Lawrence House to the center of Woodbury, then follow the signs to the Red Bank Battlefield.

Open: Saturday and Sunday 2 P.M.–4 P.M.
               or by appointment

Admission: Adults—25¢
               Children under 12—free

Postcards are for sale. The park contains numerous picnic tables and fireplaces.

INDEX

243

SUSSEX

RINGWOOD
Ringwood Manor
Skylands *

PASSAIC

RAMSEY
Old Stone House *

WAYNE
Dey Mansion
Van Ripen-Hopper House

RIDGEWOOD
Paramus School House *

MORRIS

STANHOPE *
Waterloo Village

PATERSON *
Lambert Castle

DUMONT *
Zabriskie Homestead

MORRISTOWN *
(Acorn Hall)
Ford Mansion
Macculloch Hall *
Schuyler-Hamilton House
(Speedwell Village)
Wick House

BERGEN

RIVER EDGE
Von Steuben House

CALDWELL *
Grover Cleveland Birthplace

WARREN

LIVINGSTON
Cook House *
Force House

MONTCLAIR
Crane House *

ESSEX

WEST ORANGE
* Glenmont

NEWARK *
New Jersey Historical Society

HUDSON

HUNTERDON

UNION

SPRINGFIELD *
Cannon Ball House

CLINTON *
Old Red Mill

CRANFORD *
Cranford Historical
Society

ELIZABETH *
Boxwood Hall
Belcher Mansion

SOMERVILLE *
Old Dutch Parsonage
Wallace House

PLAINFIELD *
Drake House

ANNANDALE *
Watercress Farm

MIDDLESEX

FLEMINGTON *
(Doric House)

SOMERSET

NEW BRUNSWICK
Buccleuch Mansion *

MIDDLETOWN
Marlpit Hall *

LAMBERTVILLE *
(Marshall House)

HOPEWELL *
Hopewell Museum

CRANBURY *
Cranbury Museum
Newold House

HOLMDEL *
Holmes-Hendrickson House

MONMOUTH

SHREWSBURY *
(Allen House)

WASHINGTON CROSSING STATE PARK *
McKonkey Ferry House

PRINCETON *
Bainbridge House
(Drumthwacket)
Morven
Rockingham

MERCER

TRENTON *
Old Barracks
Old Masonic Lodge
Trent House
Isaac Watson House

FREEHOLD *
Monmouth County Historical Association
(Clinton's Headquarters)

Allaire *